SPECIALTY AGRICULTURE

NIPA® GENX ELECTRONIC RESOURCES & SOLUTIONS P. LTD.
New Delhi-110 034

About the Editors

Prof. Jyoti Kachroo has remained gold medalist (two) and meritorious throughout her academic career, graduated and post graduated (Economics) from University of Jammu, Jammu. During her M.Phil. and Ph.D studies, she worked on "Critical Appraisal on New Agricultural Strategy" and "Impact of New Agricultural Strategy on the Economic and Social Conditions of Rural Women", respectively. Presently she is a Director (P&M) and Head, Division of Agricultural Economics and Agri-Business Management in SKUAST-Jammu. She has proposed MBA (ABM) and Ph.D. (ABM) programme in SKUAST-Jammu and was designated as first coordinator of the programme which is running successfully. Prof. Kachroo has successfully established Agri-Business Incubator Centre under the RVKY-RAFTAAR Scheme sponsored by Ministry of Agriculture & Farmer' Welfare, GoI. She has published about more than 100 articles which include book chapters, popular articles, review papers and about 75 research publications in referred journals.

Prof. Vikas Sharma is working as Head, Division of Soil Science at the Sher-e-Kashmir University of Agricultural Sciences and Technology of Jammu, India. He completed his doctorate from the Punjab Agricultural University and Post Doctorate from West Virginia University, USA. His Research interests include organic farming, soil carbon, soil and water conservation, land degradation, remote sensing and GIS. He has over 70 research articles, 30 book chapters and six books to his credit. He has been instrumental in establishing the Organic Farming Research Centre at SKUAST-Jammu.

Dr Sudhakar Dwivedi is Professor in the Division of Agricultural Economics and Agribusiness Management in Faculty of Agriculture at Sher- e-Kashmir University of Agricultural Sciences and Technology of Jammu, India. He received his Master's and PhD. degrees in Agricultural Economics from Dr B. R. Ambedkar University, India. He has more than twenty years of teaching experience in the field of agricultural economics. He remains advisor of more than 50 Post-Graduate students. Besides, he has been associated with national and international research projects and has more than sixty research papers published in journals of repute. He has published six books and more than thirty-five book chapters during his academic career.

Dr. S.P. Singh, Associate Professor, a visionary academician and meritorious Agro-economist. He has done M.Sc. (Agricultural Economics) from the Mahatma Gandhi Chitrakoot Gramodaya Vishwavidyalaya Chitrakoot, Satna, Madhya Pradesh and Ph.D. degree in Agricultural Economics in 2003 from the VBSP University, Jaunpur (U.P.). His more than 50 research papers/ articles have been published in reputed Journals and many research papers are under publication in ISSN journals. He has two books to his credit published by international publisher and has contributed to various practical manuals. He is working as Associate Professor (Agricultural Economics) in Division of Agricultural Economics & Agribusiness Management, SKUAST-Jammu, Chatha (J&K).

Dr. Anil Bhat, Assistant Professor, Agricultural Economics, Division of Agricultural Economics & ABM, SKUAST-Jammu possesses a meritorious academic record. He has done M.Sc. (Agril. Economics) from the Chaudhary Charan Singh University, Meerut and Ph.D. (Agril. Economics) from the Sher-e-Kashmir University of Agricultural Sciences and Technology of Jammu. He also holds the Post Graduate Diploma in Rural Development from Indira Gandhi National Open University, New Delhi and Post Graduate Diploma in Agricultural Marketing from Pondicherry University, Kalapet, Puducherry. Dr. Bhat is the recipient of certificates of "SPS measures under WTO" and "Agriculture in the WTO" from WTO, Geneva, Switzerland. He is also the recipient of Young Associate Scientist Award in the field of Agricultural Economics and has also received number of paper presentation awards. His field of expertise is production, marketing and agri-business management.

Ms. Sabbey Sharma is working as Assistant Professor in the Division of Agricultural Economics and Agribusiness Management, SKUAST- Jammu. Ms Sharma has done MBA (Agri Business Management) from SKUAST-Jammu and is now pursuing Ph.D (Business Management) with specialization in the area of Organizational Behaviour and Human Resource Management. She is a researcher, skill trainer and teacher with an experience of more than 7 years. Ms Sharma has published more than 25 research paper in the field of agribusiness management and allied sector in various reputed national and international journals. Ms Sharma is also the recipient of distinguished service award for the outstanding contribution in the field of Food Industry and Junior Young Scientist Award in the field of Agribusiness Management by Society of Biological Sciences and Rural Development.

Dr. Malika Sharma, Assistant Professor, Business Management, Division of Agricultural Economics & ABM, SKUAST-Jammu possesses a meritorious academic record. She has done MBA (Supply Chain and Marketing) and Ph.D. (Lifestyle Marketing) from Shri Mata Vaishno Devi University, Katra. She has 15 publications in Journals of high repute indexed in Australian Business Dean Council, Scopus, Web of Science and NAAS. including manuals and technical bulletins. She is the member of professional societies and her field of expertise is marketing and strategic management.

SPECIALTY AGRICULTURE

Jyoti Kachroo
Director (Planning and Monitoring)
Professor & Head, Division of Agricultural Economics and ABM
SKUAST-Jammu, Chatha – 180009, Jammu and Kashmir, India

Vikas Sharma
Professor & Head, Division of Soil Science and Agricultural Chemistry
SKUAST-Jammu, Chatha – 180009, Jammu and Kashmir, India

Sudhakar Dwivedi
Professor (Agricultural Economics)
Division of Agricultural Economics and ABM
SKUAST-Jammu, Chatha – 180009, Jammu and Kashmir, India

S.P. Singh
Associate Professor (Agricultural Economics)
Division of Agricultural Economics and ABM
SKUAST-Jammu, Chatha – 180009, Jammu and Kashmir, India

Anil Bhat
Assistant Professor (Agricultural Economics)
Division of Agricultural Economics and ABM
SKUAST-Jammu, Chatha – 180009, Jammu and Kashmir, India

Sabbey Sharma
Assistant Professor (Business Management)
Division of Agricultural Economics and ABM
SKUAST-Jammu, Chatha – 180009, Jammu and Kashmir, India

Malika Sharma
Assistant Professor (Business Management)
Division of Agricultural Economics and ABM
SKUAST-Jammu, Chatha – 180009, Jammu and Kashmir, India

nipa®

NIPA® GENX ELECTRONIC RESOURCES & SOLUTIONS P. LTD.
New Delhi-110 034

NIPA® GENX ELECTRONIC RESOURCES & SOLUTIONS P. LTD.

101,103, Vikas Surya Plaza, CU Block
L.S.C.Market, Pitam Pura, New Delhi-110 034
Ph : +91 11 27341616, 27341717, 27341718
E-mail: newindiapublishingagency@gmail.com
www: www.nipabooks.com

For customer assistance, please contact
Phone: + 91-11-27 34 17 17
Fax: + 91-11-27 34 16 16
E-Mail: feedbacks@nipabooks.com

ISBN: 978-93-95763-10-3

Composed and Designed by NIPA®.

Sher-e-Kashmir University of Agricultural Sciences and Technology of Jammu

Prof. J. P. Sharma
Ph.D. IIT Delhi
Vice Chancellor

Foreword

Agriculture is the mainstay of Indian economy as it is the primary source of livelihood for about 58% of India's population. Indian agriculture has come a long way from being a food deficit country during the time of independence to a food surplus nation in the recent years. The green, white, yellow and blue revolution paved the way for achieving this status. The due attention given to agriculture, dairy sector and other has led to remarkable growth in these sectors and has made India self-sufficient in almost all the commodities. Today, India is the largest producer of pulses, spices, milk, tea, cashew, jute, banana, jackfruit and many others in the world. However, the agriculture sector is still grappling with a lot of problems like low farmer income, shrinking of cultivable land, scarcity of water, small land holdings of majority of farmers, fragmented supply chain, huge post harvest losses, excessive use of pesticides and chemical inputs, lack of market integration, high input costs, multiple market intermediaries, lack of interest in the new generation and many others. All these are causing a strain on our environmental resources and discouraging the new generation from taking on agriculture as their profession of choice. Thus, it is imperative that new approaches are adopted which are environmentally sustainable and financially rewarding. Specialty agriculture is one such path that can be taken to alleviate some of these problems. It refers to sustainable, high-yield agriculture that can be economically sustainable even

Chatha, Jammu, J & K 180 009, INDIA
Tel.: +91-191-2263714, Fax.: +91-191-2262073
E-mail: vc@skuast.org.vcskuastjammu@gmail.com
Visit us at: www.skuast.org

in mountainous regions and small land areas. It will include, but is not limited to, the agricultural aspects of greenhouse production, hydroponics, precision farming, urban agriculture, community gardens and specialty crops like fruits and vegetables, nuts, dried fruits, greenhouse vegetable crops, horticulture and nursery crops including floriculture. The concept of Specialty Agriculture includes technological advances improving yield, and business and marketing practices supporting the small farmer. Specialty crops usually command a higher price in the market as compared to the traditional food grains and as such can be instrumental in doubling the farmers' income and improving his socio-economic condition. Crops like mushrooms, lavender, saffron, bamboo, bonsai, berries, flowers can be financially highly rewarding. They have a lot of scope of processing also which can increase their remunerative capability manifold. They can be grown in small areas and indoors as well thereby making them an attractive and profitable option for small and marginal farmers who have small land holdings and lesser resources. They shall also help in reducing the burden and strain on the environmental resources. The UT of J&K also has tremendous scope in specialty agriculture. Horticulture crops like apricots, peach, pear, apples and stone fruits, niche crops like lavender, saffron, kaala zeera, basmati, anardana and garlic along-with apiculture, aquaculture and mushroom cultivation hold a lot of potential in J&K. The production of these crops is already going on in full swing. However, the potential in processing, value addition, marketing and branding is still underutilized and presents an immense potential. The need of the hour is to develop competencies of farmers in packaging, branding and marketing which will help them in increasing their incomes and shall also lead to diversification of crops.

I congratulate the editorial team for their efforts in editing this book on 'Specialty Agriculture'. I wish them best of luck for their future endeavors also. I believe this publication shall be immensely valuable for all the stakeholders.

J. P. Sharma
Vice Chancellor

SKUAST Jammu
Chatha, Jammu, J & K 180 009, INDIA

Tel.: +91-191-2263714, Fax.: +91-191-2262073
E-mail: vc@skuast.org.vcskuastjammu@gmail.com
Visit us at: www.skuast.org

"An institution for sustainable agriculture for food and nutritional security"

Preface

The Indian agriculture sector has experienced buoyant growth in the past two years. A trend accompanying growing consumption and increasing production is the qualitative shift to special agriculture. There has been a marked shift in production from basic food grains to high value produce, especially fruits and vegetables. Specialty agriculture involves commercial agricultural operation, other than a confined feeding operation, which requires smaller tracts of land due to the intensive nature of the operation. It includes nurseries, greenhouses, mushroom farms, market gardens and apiaries. The specialty crop sector has been consistently growing over the last decade at national and international level. However, the COVID-19 pandemic has presented unique challenges to the industry that threaten its continued growth, while at the same time prompting a positive trend in consumer preferences for direct farm specialty crop purchases. Future research should examine alternative approaches to develop this important cluster of economic activity. Identifying new locations for farmers' markets, expanding sales at existing farmers' markets, and identifying opportunities for producers to sell specialty crops to new value-added markets could be among possible growth strategies.

The book contains twenty six numbers of chapters covering all the relevant topics of specialty agriculture viz., Adoption of Micro Irrigation technology: A case study of four Indian states; Organic Saffron Production: Possibilities and Approaches; Opportunities and scope of Farm Mechanization in Himalayan region; Livestock and its impact on the Environment; Importance of Bio fertilizers in Agriculture; Celery Based Crop Rotations in Punjab: Move towards Higher Returns and Farm Profitability; Profitability and Marketing Efficiency of Rabi Pea in Different Agro-Climatic Zones of Himachal Pradesh; Conservation Agriculture and Soil Health; Control of Plant Diseases Under Organic Crop Production; Types of Water Sources for Irrigation in Agriculture; Importance of Farm Mechanization in Floriculture; Floriculture: Its Prospects and Scope; Organic Farming in India: A vision towards a healthy nation; Insect pests of Rose & their management by Integrated Pest Management; Integrated Scheme for Agricultural Marketing: A Review; Zoonoses Disability Adjusted Life Year (zDALY): Indicators for measurement of impact of losses to zoonotic diseases; Impact of climate change on agriculture in the North Himalayan Region; Comparative Economic Analysis of Natural Farming: A Case Study in Una District of Himachal Pradesh; A Case Study on Comparative Economics of Natural Farming in Kangra District of Himachal Pradesh; Agripreneurs and Specialty Cultivation: The Valuable Insights; Determinants of Consumer Buying Behaviour for Organic Food Products; The status of cold-water fisheries and prospects of their utilization for poverty reduction in Jammu and Kashmir; Mustard economics of KVK beneficiaries and non-beneficiaries in Samba district of Jammu division.

This edited book on 'Specialty Agriculture' is aimed to discuss the role of Specialty agriculture in economic development of India. The book has been edited and compiled under the guidance of Prof. J. P. Sharma, Vice Chancellor, SKUAST-Jammu. We hope that the book will serve as a useful reference document on the subject. Any suggestion to further improve its contents is most welcome.

Editorial Team

Contents

Foreword vii

Preface ix

1. Adoption of Micro Irrigation Technology: A Case Study of Four Indian States 1
   Subhash Chand, Prabhat Kishore and S.K. Srivastava

2. Organic Saffron Production Possibilities and Approaches 19
   Krishna Kumar, Vishal Gupta, Gayatri Jamwal, Akash Sharma P.K. Rai, V.K.Razdan, Satish Kumar Sharma Sushil Kumar Gupta and Shahid Ahmad

3. Opportunities and Scope of Farm Mechanization in Himalayan Region 33
   Manpreet Kour, B.C. Sharma, Jyoti Kachroo and Prakarti Malakar

4. Livestock and Its Impact on the Environment 51
   Vijay Kumar Sharma, Neerja Sharma, Abhay Kumar Sinha and Saurav Gupta

5. Importance of Biofertilizers in Agriculture 57
   Saurav Gupta, Vinod Gupta, Neerja Sharma, A.K. Sinha and Vijay Kumar Sharma

6. Celery Based Crop Rotations in Punjab Move towards Higher Returns and Farm Profitability 63
   Sudeep Singh, Baljinder Kaur Sidana and Arjinder Kaur

7. Profitability and Marketing Efficiency of Rabi Pea in Different Agro-Climatic Zones of Himachal Pradesh 75
   Parul Barwal , Subhash Sharma and Diksha Bali

8. Conservation Agriculture and Soil Health 95
Abhay Kumar Sinha,Vinod Gupta, Neerja Sharma and Saurav Gupta

9. Control of Plant Diseases Under Organic Crop Production 103
Saurav Gupta,Vinod Gupta, Neerja Sharma and A.K. Sinha

10. Types of Water Sources for Irrigation in Agriculture 111
A.K. Sinha, Vinod Gupta, Neerja Sharmaand Saurav Gupta

11. Importance of Farm Mechanization in Floriculture...........................117
A.K. Sinha, Vinod Gupta, Saurav Gupta and Neerja Sharma

12. Floriculture Its Prospects and Scope...123
Neerja Sharma, Vinod Gupta, Saurav Gupta, Rakesh Sharma and A.K. Sinha

13. Organic Farming in India: A vision towards a healthy nation 131
Vinod Gupta, Neerja Sharma, Saurav Gupta and A.K. Sinha

14. Insect Pests of Rose & Their Management by Integrated Pest Management...143
Saurav Gupta, Vinod Gupta, Neerja Sharma and A.K. Sinha

15. Integrated Scheme for Agricultural Marketing A Review 157
Sheema Khan and Poonam Parihar

16. Zoonoses Disability Adjusted Life Year (zDALY): Indicators for Measurement of Impact of Losses to Zoonotic Diseases.............175
Arvind Kumar, Harsh Kumar Sharma, Rashmi Kailu Atul Anand, Zuhaib Fayaz Bhat, Sunil Kumar and Anil Bhat

17. Impact of Climate Change on Agriculture in the North Himalayan Region 187
Ankita Kumari andPavnesh Kumar

18. Comparative Economic Analysis of Natural Farming: A Case Study in Una District of Himachal Pradesh...........195
Subhash Sharma, Archita Sharma, Chinglembi Laishram and Rohit Vashisth

19 A Case Study on Comparative Economics of Natural Farming in Kangra District of Himachal Pradesh 207

Subhash Sharma, Pradeep Puria, Chinglembi Laishram and Rohit Vashisth.

20. Agripreneurs and Speciality Cultivation: The Valuable Insights 219

Shelleka Gupta, Bonia Sharma and Ronnie Dutt

21. Determinants of Consumer Buying Behaviour for Organic Food Products 233

Abid Sultan, Nikhil, Kusumakar Gautam, F.A Shaheen Arshad Bhat, Irtiqa Malik and Rehana Rashid

22. The Status of Cold-water Fisheries and Prospects of Their Utilization for Poverty Reduction in Jammu and Kashmir 245

Sahar Masud

23. Mustard Economics of KVK Beneficiaries and Non-beneficiaries in Samba District of Jammu Division 255

Tsewang Dolma, Vinod Gupta, Sudhakar Dwivedi Ruhana Rafiq and Chanchal

24. Forecasting of Trends in Production and Import of Edible Oil in India using ARIMA Model 263

P. Kanakarani and J. Gayathri

25. Status and Performance of Crop Insurance in Himachal Pradesh and India 283

Amit, Niyati Thakur, Shilpa, Ajit Sharma and Harpreet Kaur

26 Biodiversity Conservation 299

Veena Sharma Mahender Singh and Vikas Sharma

1

Adoption of Micro Irrigation Technology A Case Study of Four Indian States

Subhash Chand[1], Prabhat Kishore[2] and SK Srivastava[3]

[1]Corresponding author: Subhash Chand., Principal Scientist Agricultural Economics, ICAR-National Institute of Agricultural Economics and Policy Research, DPS Marg, Pusa, New Delhi, India
[2]Prabhat Kishore, Scientist, Agricultural Economics, ICAR-National Institute of Agricultural Economics and Policy Research, DPS Marg Pusa, New Delhi, India
[3]SK Srivastava, Sr. Scientist Agricultural Economics, ICAR-National Institute of Agricultural Economics and Policy Research, DPS Marg, Pusa New Delhi, India

Abstract

The is study analyse the impact of micro irrigation adoption and collate its outreach across the Indian states. About 14.5 million hectares (Mha) area has been brought under micro irrigation, of which sprinkler irrigation contributes 53.10% and remaining under drip irrigation system. Our estimates of micro irrigation potential for the country varied between 72.17Mha to 77.58 Mha. The Micro Irrigation implementation model adopted by Andhra Pradesh and Gujarat is found to be better and can be replicated in other states. At the field, adoption of micro irrigation helped in reducing input cost and improving farm productivity that led to appreciative net income for the farmers. The states like Andhra Pradesh and Gujarat, where dedicated implementing agencies- APMIP and GGRC, shows substantially higher adoption of micro irrigation. However, this study identifies many challenges to enlarge area under micro irrigation across the states viz. lack of institutional support (credit, accessibility, equity and awareness), irregular power supply, high initial cost, shrinking land holding size etc. Therefore, this study suggests to have state-specific schemes and implementing agencies to scale up micro-irrigation area for better trade-off between environmental sustainability and food security.

1. Introduction

Country has vast network of 20 river basins with utilizable surface and groundwater resources of 1123 billion cubic meter (BCM). With the rising population, the per capita availability of water in India has declined from 5178 m^3/year in 1951 to 1441 m^3/year in 2015. About 60% of Indian population have per capita water availability close to or lower than the water scarcity threshold of 1000 m^3/year. The per capita water availability will further decline to 1139 m^3/year with increase in population. On the demand side, the gross water requirement for all users in India is expected to grow up to 1447 BCM in 2050 (CWC, 2010, GoI, 2015). With the massive financial investment by the governments and the farmers, net irrigated area in the country increased from 20.85 million hectare (Mha) in 1950-51 to 68.38 Mha in 2014-15 (DES, 2019). Although India is a world leader in irrigation infrastructure, still half of the total cropped area (51%) remains rainfed. Further, many studies have flagged sustainability and equity concerns in irrigation development in the country (Selvarajan and Roy, 2004; Narayanamoorthy, 2011; Srivastava *et al.,* 2014). The over-dependence on groundwater sources has raised several sustainability issues and its socio-economic and ecological manifestations in many pockets of the country (Janakaranjan and Moneach, 2006; Shah, 2007; Kumar *et al.,* 2013). Motorized irrigation technologies are generally more capital-intensive than manual technologies, but they reduce drudgery and facilitate irrigation of bigger areas of land (Boutraa *et al.* 2011; Shah *et al.* 2017). Farmers level factors, e.g., age, gender and education, experience, family size, access to agricultural extension and financial services, can also affect irrigation adoption (Doss 2006, Peterman *et al.* 2014; Theis *et al.* 2018; Upadhyay 2004, Ndiritu *et al.* 2014; Salazar and Rand 2016).

Low level of water use efficiency (WUE) in agriculture is a serious challenge for sustainable development of water resources. Several demand-side management and supply-side augmentation measures have often suggested for holistic management of water resources (Rosegrant, 1997; Kumar, 2003; Briscoe and Malik, 2006, Chand S. *et al*., 2020). It is estimated that with 10% increase in present level of WUE in irrigation projects, an additional 14 Mha area could be brought under micro irrigation from the existing irrigation capacities, which would involve a very modest investment as compared to the investment that would be required for creating equivalent potential through new schemes (Swaminathan, 2006). In such circumstances, when demand for irrigation water is increasing due to promotion of water intensive agriculture, simple and accessibility to technology, energy and funds. The micro irrigation is the answer to such questions and can be the economical proposition for the future agriculture. Namara *et al.* (2007), in their studies in Maharashtra and Gujarat state of India found significantly positive effect of ownership of well and high horsepower pumps on likelihood of adaptation of micro-irrigation technologies. Literature shows many energy efficient interventions of micro irrigation i.e. bucket kits (Fandika *et al.*, 2012), IDE low-cost drip irrigation (Polak *et al.*, 1997; Van Westarp *et al.*, 2004), Drum kit (Karlberg *et al.*, 2007; Kulecho and Weatherhead, 2005), gravity-fed micro irrigation (Bhatnagar and Srivastava, 2003; Kumar *et al.*, 2009) and energy system (Verma *et al.*, 2004). Thus, the adoption of micro irrigation technologies may not automatically lead to water saving at the basin level, unless enabling institutional and economic policy instruments are put in place that allow the equitable distribution or allocation of the saved water. In the backdrop of this, the paper analyses the adoption of micro irrigation and its impact, constraints of adoption of micro irrigation in India, in order to gain insights for upscaling this technology.

2. Materials and methods

Selection of study area and sampling design: Purposively four states namely Punjab, Andhra Pradesh, Gujarat and Maharashtra were selected for study. The condition was three states where micro irrigation spread is very high and other state i.e. Punjab where MIS spread is low but has high potential for increasing the coverage under micro irrigation. A stratified random sampling technique was followed for selection of respondents. Based on the secondary data all districts of stated states were arranged in increasing order of area under micro irrigation system (MIS). From each state, one district with higher area and one district with low coverageof MIS was selected for primary survey. Further two blocks from each district were selected randomly. Selection of villages and respondent wasattempted randomly. Sample consisting of micro irrigation adopter and non-adopter from each state were targeted for intensive survey.

Data and data analysis: Primary and secondary data have been collected for the study. The primary information was collected through personal interview method on pre-tested questionnaire. Based on sampling design, detailed survey of 827 respondents was carried out which comprised of almost equal number of adopter and non-adopter. Primary data includes socio economic profile of adopter and non-adopters, crop grown, source of irrigation, year of MIS installation, cost structure, farmers perception on MIS use, operational procedure followed by the beneficiaries, cost of crop cultivation with and without MIS, source of information for micro irrigation etc.The secondary information were collected from the line departments of selected states regarding the extent of area under MIS, operational procedure adopted by the concern department involved in its implementation, cost and subsidy norms, selected beneficiaries, irrigation source etc. The PKMSY guideline was accessed from government website. The series of discussion and focus group meeting were conducted with micro irrigation implementing agencies at field level, farmers and martial supplying firms to understand the procedure followed. We have analyzed the strengths and weakness (SWOT) of the operational and administrative procedure adopted while implementing the micro irrigation scheme. To quantify the perceptions of farmers on scope of opportunities developed from the development of MI, we have used Likert type scale of five-point (Very high=5, High=4, Neutral=3, Less scope=2, No scope =1).

Analytical frameworks: Average yield, water productivity and technical efficiency for wheat growers using sprinklers were compared with their counterparts following flood method of irrigation. While this is useful to demonstrate efficiency gains from sprinkler irrigation, a simple comparison of means is a biased measure of gains because adoption of sprinkler involves a selection process. Farmers may self-select to adopt sprinkler irrigation, or they may be selected as beneficiaries of DBT scheme. Hence it is likely that population of adopters differs from that of non-adopters.

To take this into account, we adopt treatment effects models from the programme evaluation literature (Two stage Heckman procedure). In a regression framework, the treatment effects model is given by:

$$R_i = a + bC_i + c' X_i + \varepsilon_i \qquad (1)$$

Where, is an outcome variable (yield, water productivity, technical efficiency) for farmer *i*, is a dummy variable taking value 1 for a farmer who has adopted sprinkler method of irrigation and otherwise 0. is a vector of control variables, and is zero mean random variable.

An ordinary least squares estimate of Eq.(4) is likely to be biased if ε_i contains within it random unobservable factors such as ability that are not uniformly distributed within the population of adopter and non-adopters. In such a case, error term is likely to be correlated with C_i. Thus, for instance, if adopters are more productive than non-adopters because of unobserved ability, then a simple comparison of the means as well the OLS estimates of Eq.(1) would yield an overestimate of the true measure of gains from adoption. Hence, we apply two-stage Heckman procedure to correct for the bias from the endogeneity of right hand side variables.

Consider the following adoption equation:

$$C_i = \alpha_i Z_i + u_i \qquad (2)$$

Where, C_i is a binary variable (1 for adopters and 0 for non-adopters), Z_i is a vector of variables that matters for adoption, Variables in Z_i will overlap with variables in X_i. Identification requires that there be at least one variable in Z_i that is not in X_i. If this condition is met, the predicted value from (5) can be used as instrument for C_i in regression equation (4). Thus, from Eq.(5) we estimate inverse mill ratio (IMR) and use as an instrument in Eq.(4). This would yield a consistent estimate of b provided the instruments are uncorrelated with the error term in Eq.(2).

3. Results and Discussion

Evolutions and implementation of Micro irrigation scheme in India

The Journey of Micro irrigation (MI) in India was started in 1981 when the National Committee on Plasticulture in Agriculture (NCPA) approved the use of plastics in agriculture on a pilot basis. The NCPA, in its four successive report (1982, 1983, 1984, and 1985), emphasized upon promoting use of plastics for drip irrigation, mulching and green houses to boost horticulture production. Based on the recommendations of NCPA, the Government of India launched a centrally-sponsored scheme on the use of plastics in agriculture in 1992. Under the scheme, farmers were eligible for financial assistance or subsidy for installing MI system depending on land size, cost and economic condition of farmers. To accelerate spread of MI technologies, the Government created the Rural Infrastructure Development Fund (RIDF) under the purview of National Bank for Agriculture and Rural Development (NABARD) in 1995-96 with an initial corpus of Rs. 2000 crore. A total of 36 eligible activities including micro, major and medium irrigation projects were provided the financial assistance through RIDF. In 1996-97, The Extension, Renovation and Modernization (ERM) of irrigation projects, with the provision of implementation of MI in at least 10% of command area, were given priority in extending financial support. Similarly, in the centrally-sponsored Integrated Scheme of Oilseeds, Pulses, Oil Palm and Maize (ISOPOM) launched in 2004 in 14 major oilseeds growing states, financial assistance was provided for creating efficient irrigation infrastructure through distribution of sprinkler sets or drip system. The launching of National Horticulture Mission (NHM) in 2005 strengthened the MI activities and accelerated its spread on a large scale. With broad objectives for the holistic growth of horticulture sector, NHM was restructured into a Mission for Integrated Development of Horticulture (MIDH) in 2014-15. In MIDH, adoption of MI is being fostered through the activities of protected cultivation and creation of water sources.

Having realized the potential benefits of MI technology in conserving water resources and sustaining crop yield, the Government of India launched a dedicated scheme named Centrally-Sponsored Scheme (CSS) on MI on January 20, 2006. The main objective of the scheme was to enhance WUE in agriculture sector by encouraging farmers to adopt appropriate technological interventions like drip and sprinkler irrigation. In the flagship scheme, Rashtriya Krishi Vikas Yojana (2007), MI was included as one of the components for ensuring an aggregate growth rate of 4% in agriculture and allied sector. To bring all states, including north eastern and Himalayan states, under the ambit of MI scheme, the Government upgraded the CSS on MI into National Mission on Micro irrigation (NMMI) in June 2010 and further to National Mission on Sustainable Agriculture (NMSA) in April 2014 and implemented On Farm Water Management (OFWM) programme from year 2014-15

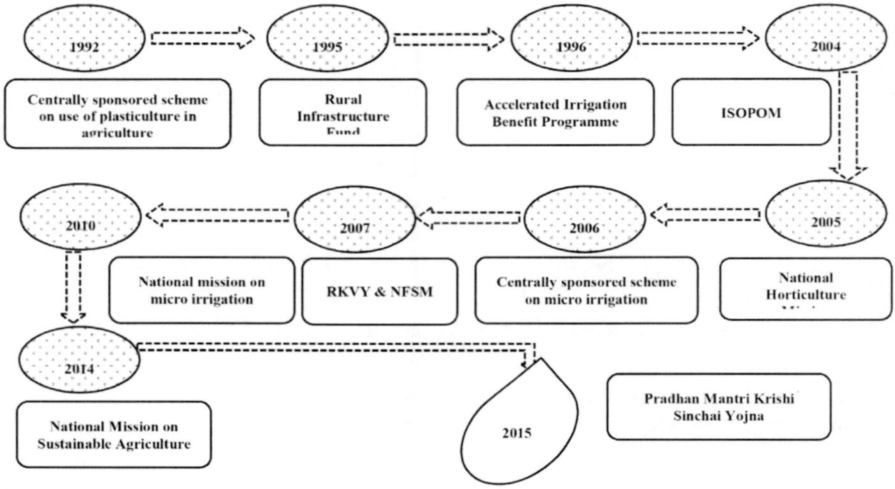

Fig. 1. Evolution of micro irrigation scheme in India
Source: Prepared by authors Architecture of PMKSY

. From the year 2015-16, the Government has subsumed all existing schemes of irrigation into Pradhan Mantri Krishi Sinchayee Yojana (PMKSY). The main motive of PMKSY is to provide water to every field (*har khet ko pani*), improve on-farm water use efficiency, enhance adoption of precision irrigation and water-saving technologies (*per drop more crop*). An outlay of Rs. 50,000 crores over a period of five years (2015-16 to 2019-20) was allocated for PMKSY. A timeline of development of schemes promoting MI in the country is presented in Figure 2.

Current status of micro irrigation coverage in Indian states

With concerted efforts of Indian government, 14.71 Mha cumulative area has been brought under micro-irrigation across states up to 2021-22. Ambitious target of 10 million hectares was approved under *Pradhan Mantri Krishi Sinchayee Yojana* (PMKSY) for a period of 5 years (2015-16 to 2019-20). Since then, an area of 4.44 million hectares has been brought under micro-irrigation, sprinkler and drip irrigation contributed 46% and 54% respectively (PMKSY, 2020). At present, five states namely Rajasthan, Andhra Pradesh, Maharashtra, Karnataka and Gujarat contribute 3/4th cumulative area covered under micro-irrigation. Distribution of micro-irrigation area across states has been depictedinfig1. The states could not make significant progress in micro irrigation coverage need to speed up the efforts so that quantum of input saving can improve the income of the famers in the respective states.

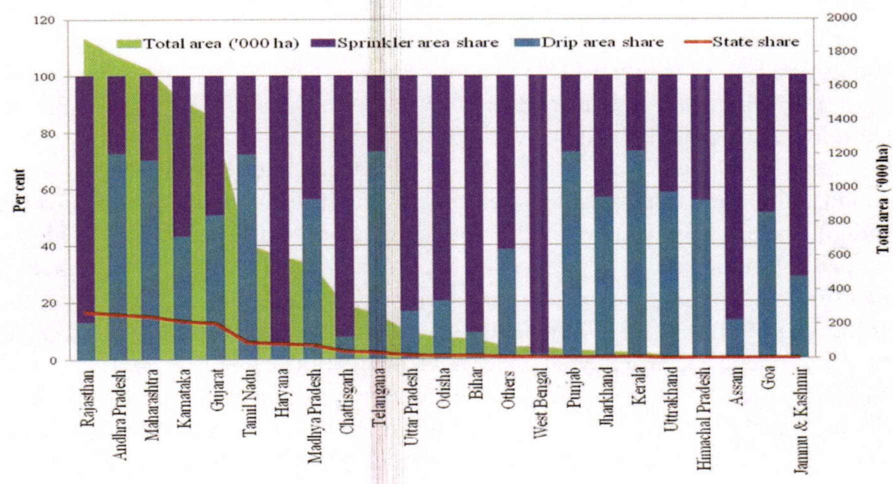

Fig. 2. Distribution of micro-irrigation area across states
Source: GoI, 2019.

Estimation of micro irrigation potential across Indian states: We have classified the states based on micro irrigation penetration and same is presented in table 1. The classification has of states is based on the criteria, having micro irrigation coverage >20% of total irrigated areas kept in A category. Similarly, those states having micro irrigation coverage 10% to 20% has been kept in B category and remaining in C category. However, government of India guidelines suggests different distribution of all Indian states and based on this classification government of India allocate the subsidy and funds to the state government. Thus, our analysis of the Indian state may provide policy direction to the government policy makers to decide the policy frameworks for scaling up the micro irrigation in the country.

In our analysis, estimate of potential area for micro-irrigation came out to be about 72 and 78 million hectares with index 1 and index 2 respectively. In scenario 1, sprinkler irrigation potential estimated to be 51.01 million hectares which is 2/3rd of total potential area estimated and rest is contributed by drip irrigation. Of the total potential estimated under scenario 1, Uttar Pradesh shares largest potential area (25%) followed by Rajasthan (12%) and Madhya Pradesh (10%). In scenario 2, absolute potential area has increased by 7.44% compared to scenario 1 potential estimate. If all estimate compared with the present area achieved under micro irrigation, then too there are much variation in potential existing and realised. For instance, sprinkler irrigation, variation in area achieved to its potential estimates ranges from 10.6 to 17.8 per cent whereas for drip irrigation it ranges from 17.8 per cent to 41.2 per cent.

The Rajasthan and Karnataka state achieved highest area coverage through sprinkler irrigation to its micro irrigation potential followed by Chhattisgarh and Andhra Pradesh (Annexure-I&II). Andhra Pradesh which contributes nearly 1/4th of total drip irrigated area in country, have covered about 93% of its estimated micro irrigation potential. However, in other states much emphasis is on sprinkler irrigation for vegetables and other crops. Maharashtra and Jharkhand state ranked second and third in drip irrigation area coverage to its estimated potential. Overall, Andhra Pradesh achieved nearly 2/3rd of estimated potential area which is highest compared to other states. This may be due to collective effort of government in creation of Special Purpose departments namely Andhra Pradesh Micro-Irrigation Project (APMIP) in 2003. With scenario 1, at country level, 15.81% of estimated potential area has been achieved under micro-irrigation (table 1, *Scenario 1*). Total estimated potential increased from 72.17 to 77.58 million hectare due to addition of more canal irrigated area (*Scenario 2*). So, estimate of Potential , may show lower penetration as more area has come under potential estimate of each states. Overall, highest penetration of micro-irrigation has been made in Andhra Pradesh followed by Karnataka and Maharashtra. On basis of this estimate, 14.71% of estimated area of India has been covered under micro-irrigation.

Table 1: Estimates of micro-irrigation potential area and its outreach across Indian states

States	Ultimate irrigation potential (MHa.)	Potential estimates (MHa.)		Penetration of MIS % to Potential area	
		Scenario-1	Scenario -2	Scenario-1	Scenario -2
High level of micro irrigation penetration (> 20%)					
Andhra Pradesh	3.67	2.62	2.88	68.08	61.91
Karnataka	3.83	2.46	2.72	61.81	55.87
Maharashtra	4.26	3.27	3.55	52.18	48.01
Chhattisgarh	1.8	0.93	1.15	34.01	27.56
Gujarat	6.05	5.12	5.34	27.78	26.63
Tamil Nadu	3.27	2.71	2.87	24.9	23.54
Jharkhand	0.23	0.17	0.17	21.92	21.74
Rajasthan	10.49	8.46	8.99	22.36	21.04
Medium level of micro irrigation penetration (10% to 20%)					
Haryana	0.21	0.05	0.05	20.25	19.63
Odisha	1.41	0.62	0.85	19.78	14.51
Himachal Pradesh	5.83	4.15	4.63	14.6	13.08
Kerala	0.47	0.27	0.29	12.12	11.29
Telangana	2.57	2.28	2.35	11.5	11.18
Low level of micro irrigation penetration (<10%)					
Madhya Pradesh	10.08	7.49	7.83	7.43	7.11
Goa	0.04	0.03	0.03	7.44	7.06

States	Ultimate irrigation potential (MHa.)	Potential estimates (MHa.)		Penetration of MIS % to Potential area	
		Scenario-1	Scenario -2	Scenario-1	Scenario -2
Uttarakhand	0.54	0.41	0.44	2.97	2.77
Bihar	5.29	3.93	4.26	2.94	2.71
Assam	0.41	0.13	0.15	2.1	1.82
West Bengal	5.49	3	3.71	2.22	1.8
Uttar Pradesh	20.52	17.65	18.41	0.87	0.84
Punjab	7.75	6.27	6.69	0.78	0.73
Jammu & Kashmir	0.48	0.15	0.24	0.05	0.03
India	94.68	72.17	77.58	15.81	14.71

Source: Author's calculation as Scenario 1 & 2 estimated based on irrigation index 1& 2 respectively.

Administrative and operational constraints in micro irrigation

The sample farm households of study states of Punjab, Andhra Pradesh, Gujarat and Maharashtra felt that the process of getting MI is lengthy. This can be smoothened through administrative reforms in the line departments involved in promotion of MI. Where farmers and line departments had better understanding and cooperation, process of MI access was simple. In some areas of Punjab, adopters reported favourism based on the caste, association with the political party, and status in the society. A small number of respondents in each state also informed about prevalence of nepotism. During post-installation period, farmers faced several constraints like clogging of laterals and pipes. Although concerned line departments had clear instructions of maintenance by installing firms, still a large number of farmers faced problems of clogging and chocking of laterals.

In Andhra Pradesh, Gujarat and Maharashtra, programme implementing agency (PIA) and firms organized a series of training programs for farmers and field level staff on maintenance. In Punjab, availability of post-installation service was perceived to be poor as compared to other states. The damage to MI by own or wild animals was noticed in all the states. Farmers have reported that once laterals are damaged, it is very difficult to replace or to get serviced due to remote location of the fields. In the state of Maharashtra and Gujarat, there were incidences of theft of MI system. Difficulties in maintenance and unreliable energy source were other constraints faced by the respondents. For scaling up of the MI, these constraints need to be addressed on priority.

Adoption of Micro Irrigation: This section discusses the benefits realized by the farm households adopting MI technologies. The benefits were estimated using farm-level data on cost of cultivation collected through farm survey from both adopter and non-adopter farmers of Punjab, AP, Gujarat and Maharashtra. For the analysis, crops were selected based on their dominance in the cropping pattern of the selected states. Cost estimation includes both variable and fixed costs.

Inputs saving and productivity improvement through micro irrigation adoption: The evidences show reduction in input cost and increase in productivity and income from adoption of micro irrigation. Kapur *et al.*, 2015 Narayanamoorthi, 2018, Reddy *et al.*, 2017, Priyan and Panchal, 2017, Chandran and Surendran, 2016, Tiwari *et al.*, 2014, Kumar *et al.*, 2016 and others have worked out direct impact of micro irrigation in different states of India and our finding in conformity of these authors as given in this section. The adopter farmers have saved cost on the seed and planting material to the extent of 2.25% to 89.40% as compared to non-adopters (Table 2). The savings in cost on farmyard manure (FYM) varied from 3.17% to 57.94% in different crops whereas saving in cost from fertilizers on adopted farms ranged from 12.89% to 68.76%. It is from the fact that adopter farmers usually have applied liquid fertilizers and this way less fertilizer required when it applied through MIS. Similarly, savings in cost on chemicals and pesticides use varied from -0.92% to 90.29% across different crops in different states.

Table 2. Saving in input cost, increase in yield and income due to adoption of micro irrigation

(Percent)

States	Crops	Input savings								Output increased	
		Seed/Plants	FYM	Ferti-lizer	Chemical	Irrigation	Labour	Machine use	Total cost	Yield	Net Return
Punjab	Cotton	-2.25	-5.46	-12.89	-48.23	-14.77	-5.14	4.69	-1.33	11.62	32.27
Gujarat	Cotton	-6.67	-13.73	-53.91	-57.72	-12.6	-21.48	-8.31	-26.65	21.72	42.8
Maharashtra	Cotton	-32.27	-2.94	-15.04	-50	-27.5	-51.15	-35.81	-35.15	18.64	35.26
Punjab	Kinnow	13.48	-3.17	-18.50	-17.71	-60.24	-12.35	-17.29	-8.14	10.76	34.4
	Maize	-29.76	-9.78	-20.73	22.11	-35.21	-31.72	-14.13	-13.07	8.17	54.1
Maharashtra	Maize	19.65	-	-16.41	-	-16.43	-19.92	-12.67	-7.76	15.56	39.35
Gujarat	Bajra	-59.79	-	-45.45	-33.68	-39.46	-34.66	-9.88	-4.15	4.82	41.64
Maharashtra	Bajra	-9.21	-20.45	-22.37	-	-42.14	-12.94	-47.47	-31.31	17.37	58.69
Gujarat	Soybean	-21.28	-22.53	-63.89	-90.29	-27.48	-48.65	-20.25	-29.59	13.09	52.01
Maharashtra	Soybean	-4.91	-8.33	-36.1	-38.84	-31.13	-25.88	-35.61	-24.08	8.69	25.88
Punjab	Wheat	-46.87	3.67	-37.51	-45.30	-32.00	-13.77	-9.83	-18.06	12.96	35.14
Andhra Pradesh	Brinjal	-89.40	-16.10	-68.76	-22.62	-34.63	-22.78	9.08	-13.63	10.54	32.4
	Coconut	4.43	-7.42	-13.60	-6.25	-29.62	-26.16	-5.08	-8.52	20.17	42.03
	Papaya	34.87	6.53	-38.12	-12.48	-51.19	-26.56	-5.35	-5.31	10.51	12.28
	Tomato	3.90	-7.65	-26.38	-8.89	-16.07	3.81	1.40	-4.59	22.27	43.02
	Sugarcane	-8.14	-30.40	-21.66	-0.92	-43.59	-29.41	-25.35	-24.60	6.82	21.32
Maharashtra	Sugarcane	-11.59	-96.01	-52.09	-43.66	-28.33	-10.05	-4.22	-17.1	10.01	25.74
Gujarat	Groundn-ut	8.76	-57.94	-49.76	-34.54	-88.62	-36.03	-20.84	-30.39	10	17.63
	Potato	-4.32	-29.21	-56.38	-33.2	-37.21	-21.33	-10.64	-10.35	18.49	37.86
Maharashtra	Onion	-36.31	-10.85	-12.98	-5.08	-85.81	-8.25	-7.64	-12.74	12.54	20.95

Source: Authors' estimates Note: Total cost includes the variable and fixed cost

The savings in water used for cultivation ranged between 12.60% and 88.62% across crops grown by sample adopter households. There was a saving in labour use and machine hours used by adopters and it varied from 5.14% to 48.65% across crops selected for analysis. The net return was higher on adopted farms to the extent of 12.28% to 54.10% (Similar observation made by Gorain, *et al.*, 2018 and Narayanamoorty, *et al.*, 2018). Thus, it is clear that adoption of micro irrigation technology helps in lowering input use thereby reduction in cost, which helped in improving income of farm households. Apart from savings in input, micro irrigation helps in providing employment, reduction in migration, youths attraction toward agriculture and thereby improvement of livelihood of the farmers.

Determinants of adoption of Sprinkler irrigation and its effects

Key factors that influence adoption of sprinkler irrigation are landholding size, ownership of tube-well and electric pump, and households' association with any social organization. The probability of adoption of sprinkler irrigation is higher for larger farmers, and those who own their own tube-wells and electric engines for pumping groundwater for irrigation. Interestingly, a household's association with social organization increases chances of adoption of sprinkler irrigation perhaps due to their better awareness about benefits of sprinkler irrigation and flow of information on government schemes (Table 3).

Table 3. Determinants of adoption of sprinkler irrigation

Dependent variable: Adoption of sprinkler (yes=1; no=0)		
Independent variables	Coefficient	Marginal effect
Ln landholding size (acre)	0.617***(0.145)	0.162***(0.036)
Ln farming experience (years)	-0.261 (0.185)	-0.068(0.048)
Ln schooling (years)	0.074 (0.049)	0.019 (0.013)
Member of social organization (yes=1, no=0)	0.477**(0.233)	0.125**(0.060)
Possess a Kisan Credit Card (yes=1, no=0)	0.134 (0.194)	0.035 (0.051)
Caste (SC/ST and OBC=1, others =0)	0.185 (0.266)	0.048 (0.070)
Own tube-well (yes=1, no=0)	0.501* (0.288)	0.131*(0.075)
Own electric pump (yes=1, no=0)	0.851***(0.254)	0.223***(0.064)
Own diesel engine (yes=1, no=0)	-0.109 (0.208)	-0.029 (0.054)
Constant	-2.192***(0.804)	
Number of observations	403	
LR chi^2	65.84	
Prob>chi^2	0.0082	

Note: Figures in parenthesis are standard errors;*, ** and *** are significant at 10, 5 and 1% level respectively.

5. Conclusions and policy recommendations

After the launch of MI scheme in January 2006, the governments, farmers, and development agencies are making all efforts to increase coverage under MI thereby enhancing WUE at farm level. Presently, about 20% of estimated potential (72.2 Mha) has been covered under MI. The large attention is being given to cover more area under horticultural, other high-value and water-intensive crops like sugarcane, rice, etc. The water deficit regions are leading in the adoption of MI technologies and using this technology to mitigate water scarcity and increasing WUE. Substantial rates of returns from drip (48-135%) and from sprinkler (44-144%) irrigations including the maximum payback period of 2-3 years have also been reported (GoI., 2014). Besides, the adoption of MI helps in raising irrigated area, diversifying area towards high-value crops, enhancing yield and quality produce, and reduce cultivation cost per ha.Realizing the large benefits using MI technologies and fast depleting groundwater resources across states, concerted policy interventions are needed to increase MI coverage, particularly in slow promoting states. In this regard, strict vigil from the nodal agency on companies installing MI, supplying components and post installation repair and maintenance will bring more effective results. Promotion of solarisation programme with MI system and ongoing development in solar technology will go a long way in dealing with MI coverage. Besides, collective approach in drilling tube well and recognizing this in providing subsidy for MI will further speed up the MI adoption. Therefore, more efforts from government side should be made to speed up the coverage, particularly in states having large potential like Uttar Pradesh, Madhya Pradesh, Bihar and West Bengal, so that the significant proportion of potential area could be covered in reasonable period of time. The increase in net returns was higher on adopted farms, definitely is providing incentives for socio economic upliftment of farmers. Besides, adoption of MI provides better employment and income generation opportunities to rural youths by attracting them towards agriculture and improving their skills through capacity building programmes. Despite several benefits of micro irrigation and supports for installation in terms of subsidies available but still micro irrigation spread is not presenting the encouraging picture for India. It may be due to, several challenge in adoption of micro irrigation. Therefore, the awareness and mass contact programs should be a continuous process, so that more farmers can be brought in ambit of micro irrigation in the country.

Acknowledgement: This paper is drawn from the study funded by NITI Aayog, New Delhi.

References

Anbumozhi, V., Matsumoto, K., Yamaji, E., 2001. Towards improved performance of irrigation tanks in semi-arid regions of India: modernization opportunities and challenges. Irrig. Drain. Syst. 15 (4), 293–309.

Bhatnagar, P.R., Srivastva, R.C., 2003. Gravity-fed drip irrigation system for hilly terraces of the north-west Himalayas. Irrig. Sci. 2 (4), 151–157.

Biswas, S.K., Akanda, A.R., Rahman, M.S. and Hossain, M.A. 2015. Effect of drip irrigation and mulching on yield, water-use efficiency and economics of tomato. Plant, Soil and Environment,61 (3): 97-102.

Boutraa, T, A Akhkha, A Alshuaibi and R Atta (2011). Evaluation of the effectiveness ofan automated irrigation system using wheat crops. Agriculture and Biology Journal of North America, 2(1), 80–88. https://doi.org/10.5251/abjna.2011.2.1.80.88.

Briscoe, J. and Malik, R.P.S., 2006 India's water economy: Bracing for a turbulent future. Oxford University Press, New Delhi.

Camp, C.R., Sadler, E.J., Busscher, W.J., 1997. A comparison of uniformity measures for drip irrigation systems. Trans. ASAE 40 (4), 1013–1020.

Chand, S., Kishore, P. and Srivastava, S.K., 2019 Pressurized irrigation system: Policies and implications in India. Indian Journal of Soil Conservation, Conference, pp: 42-50.

Chand,S., Prabhat Kishore, S Kumar and Srivastava, S.K., 2020. Potential, Adoption and Impact of Micro Irrigation in Indian Agriculture in India, Policy paper No. 36, Published by ICAR-NIAP, Pusa, New Delhi.

Chandrakanth, M.G., Priyanka, C.N., Mamatha, P. and Patil, K.K., 2013 Economic Benefits from Micro irrigation for Dry Land Crops in Karnataka. Indian Journal of Agricultural Economics, 68 (3):426-338.

Chandran, M.K. and Surendran, U., 2016 Study on factors influencing the adoption of drip irrigation by farmers in humid tropical Kerala, India. International Journal of Plant Production, 10 (3): 347-364.

CWC (2010). Report of the Standing Sub-Committee on Assessment of Availability & Requirement of Water for Diverse uses-2000. XI Plan Document. Basin Planning Directorate. Central Water Commission, Government of India.

Department of Fertilizer 2019 Direct Benefit Transfer Division, Ministry of Chemicals and Fertilizer, Government of India.

DES 2019 Agricultural Statistics at a Glance, Directorate of Economics and Statistics, Ministry of Agriculture and Farmers Welfare, Government of India.

Doss, CR (2006). Analyzing technology adoption using microstudies:limitations, chal-lenges, and opportunities for improvement. Agricultural Economics, 34(3), 207–219.http://onlinelibrary.wiley.com/doi/10.1111/j.1574-0864.2006.00119.x/full.

Fandika, I.R., Pakeni, D.K., Zingore, S., 2012. Performance of bucket drip irrigation powered by treadle pump on tomato and maize/bean productivity in Malawi.Irrig. Sci. 30, 57–68.

GoAP (2017). Annual Report, Andhra Pradesh Micro Irrigation Project. Government of Andhra Pradesh, India.

GoI (2017). Operational Guidelines of Per Drop More Crop (Micro irrigation) Component of PMKSY. Ministry of Agriculture & Farmers Welfare, Government of India, New Delhi.

GoI. (2015). Water and Related Statistics, Water Resource Information Directorate, Central Water Commission, Government of India. http://www.cwc.gov.in/sites/default/files/water-related-statistics-2015.pdf.

Gorain, S., D.R. Singh, P. Kumar, P. Venkatesh and G.K. Jha (2018). Social Costs and Benefits Analysis of Drip Irrigation System in Northern Maharashtra. Economic Affairs, 63 (4):1061-1065.

Gujarat Green Revolution Company (GGRC), 2016. 19th Annual Report, Government of Gujarat, India.

Indian National Committee on Irrigation and Drainage, 1994. Drip irrigation in India, New Delhi.

Janakarajan, S. and Moneach, M. (2006). Are wells a potential threat to farmers well-being? Case of Deteriorating Groundwater Irrigation in Tamil Nadu, India. Economic and Political Weekly, 77: 3977-3987.

Jha, G., Choudhary, O. P. and Sharda, R. (2017). Comparative effects of saline water on yield and quality of potato under drip and furrow irrigation. Cogent Food & Agriculture, 3: 1369345.

Kang, Y., Nishiyama, S., 1996. Analysis of micro irrigation systems using a lateral discharge equation. Trans. ASAE 39 (3), 921–929.

Kang, Y., Yuan, B.Z., Nishiyama, S., 1999. Design of micro irrigation laterals at mini-mum cost. Irrig. Sci. 18 (3), 125–133.

Kapur, R., Gulati, S. and Chouhan, S. (2015). Accelerating growth of Indian agriculture: Micro irrigation an efficient solution. Federation of Indian Chambers of Commerce & Industry (FICCI), Strategy Paper - Future Prospects of Micro Irrigation in India.

Karlberg, L., Rockstrom, J., John, G., Annandele, J., Steyn, M., 2007. Low cost drip irrigation suitable technologies for southern Africa? An example with tomatoes using saline irrigation water. Agric. Water Manage. 89: 59–70.

Keller, J., Bliesner, R.D., 1990. Sprinkler and Trickle Irrigation. Chapman and Hall,New York.

Kulecho, I.K., Weatherhead, E.K., 2005. Reasons for small holder's farmers discontinuing with low cost micro-irrigation: a case study from Kenya. Irrig. Drain. Syst.19 (2), 179–188.

Kumar, D. (2003). Demand management in the face of growing water scarcity and conflicts in India: Institutional and policy alternatives for future. In: Water Resources and Sustainable Livelihoods and Eco-System Services, Eds: K. Chopra, C.H.H. Rao and R. Sengupta. Concept Publishing Company, New Delhi, 97-131.

Kumar, M., Kumar, N., Singh, K.P., Kumar, P., Srinivas, K., Srivastva, A.K., 2009. Integrating water harvesting and gravity-fed micro-irrigation system for efficient water management in terraced land for growing vegetables. Biosyst. Eng. 102 (1), 106–113.

Kumar, M.D., Reddy, V.R., Moorthy, A.N., and Sivamohan, M.V.K. (2013). Analysis of India's Minor Irrigation Statistics: Faulty Analysis, Wrong Inferences. Economic and Political Weekly, 48 (45 & 46):76-78.

Kumar, R., Trivedi, H., Yadav, R., Das, B. and Bist, A.S. (2016). Effect of drip irrigation on yield and water use efficiency on Brinjal (Solanum Melongena) Cv. Pant Samrat. International Journal of Engineering Sciences & Research Technology, 5 (10): 7-17.

Molden, D.; Sakthivadivel, R.; Habib, Z. (2001). Basin-level use and productivity of water: Examples from South Asia. IWMI Research Report 49, Colombo: International Water Management Institute.

Narayanamoorthy A. (2011). Development and composition of irrigation in India: temporal trends and regional patterns. Irrigation and Drainage, 60: 431–445.

Narayanamoorthy, A. (1997). Drip irrigation: A viable option for future irrigation development. Productivity, 38 (3): 504-511.

Narayanamoorthy, A. (2003). Averting water crisis by drip method of irrigation: A study of two water-intensive crops. Indian Journal of Agricultural Economics, 58 (3):427.

Narayanamoorthy, A., Bhattarai, M., and Jothi, P. (2018). An assessment of the economic impact of drip irrigation in vegetable production in India. Agricultural Economics Research Review, 31 (1), 105-112.

Namara, R.E., R.K. Nagar and B. Upadhyay (2007) Economics, Adoption Determinants, and Impacts of Micro-irrigation Technologies: Empirical Results from India. Irrigation Science, 25: 283-297, DOI 10.1007/s00271-007-0065-0

NCPAH (2014). National Committee on Plasticulture Application in Horticulture. Ministry of Agriculture & Farmers Welfare, Government of India.

Ndiritu, SW, M Kassie and B Shiferaw (2014). Are there systematic gender differences inthe adoption of sustainable agricultural intensification practices? Evidence fromKenya. Food Policy, 49, 117–127. https://doi.org/10.1016/j.foodpol.2014.06.010.

Panigrahi, B., Roy, D.P. and Panda, S. N. (2015). Water use and yield response of tomato as influenced by drip and furrow irrigation. International Agricultural Engineering Journal, 19 (1): 19-30.

Paul, J.C., Mishra, J.N., Pradhan, P.L. and Panigrahi, B. (2013). Effect of drip and surface irrigation on yield, water-use-efficiency and economics of capsicum grown under mulch and non-mulch conditions in eastern coastal India.European Journal of Sustainable Development, 2 (1) 99-108.

Peterman, A, JA Behrman and AR Quisumbing (2014). A review of empirical evidenceon gender differences in nonland agricultural inputs, technology, and servicesin developing countries.Gender in Agriculture: Closing the Knowledge Gap,11,145–186. https://doi.org/10.1007/978-94-017-8616-47.

Polak, P., Nanes, B., Adhikari, D., 1997. A low cost drip irrigation system for small farmers in developing countries. J. Am. Water Res. Assoc. 33 (1), 119–124.

Raina, J.N., Sharma, T. and Suman, S. (2011). Effect of drip fertigation with different fertilizers on nutrient distribution in soil, leaf nutrient content and yield of apricot (Prunus aremeniaca L.). Journal of the Indian Society of Soil Science, 59: 268-77.

Rao, K.V.R., Gangwar, S., Keshri, R., Chourasia, L., Bajpai, A., and Soni, K. (2017). Effects of Drip Irrigation System for enhancing Rice (Oryza Sativa L.) yield under system of rice intensification management. Applied Ecology and Environmental Research,15 (4):487-495.

Rosegrant, M. (1997). Water resources in 21st century: Challenges and implications for action. Food, Agriculture and Environment Discussion Paper 20: International Food Policy Research Institute, Washington DC.

Salazar, C and J Rand (2016). Production risk and adoption of irrigation technology:Evidence from small-scale farmers in Chile.Latin American Economic Review,25(1). https://doi.org/10.1007/s40503-016-0032-3.

Selvarajan, S. and Roy, B.C. (2004). Water–food security analysis. In Proceedings of Planning Methodology Workshop on Resource Analysis for Sustaining Water–Food Security, National Centre for Agricultural Economics and Policy Research (NCAP), New Delhi, India, pp:1–16.

Shah, T. (2007). The groundwater economy of South Asia: An assessment of size, significance and socio-ecological impacts, IWMI-TATA Water Policy Program, Field Office, Anand, Gujarat, India.

Sheahan, M., and CB Barrett (2017). Ten striking facts about agricultural input use in Sub-Saharan Africa. Food Policy, 67:12–25. https://doi.org/10.1016/j.foodpol.2016.09.010.

Sivanappan, R.K., 1994. Prospects of micro-irrigation in India. Irrig. Drain. Syst. 8 (1),49–58.

Upadhyay, B., 2003. Drip irrigation: an appropriate technology for women. Approp.Technol. 30 (4), 31–37.

SPACE (2021) Study on Efficacy of Micro-Irrigation System in Drought Prone Parts of Haryana. NABARD Research Study 15, completed by Society for Promotion and Conservation of Environment (SPACE), Chandigarh.

Srivastava, S.K., Ghosh, S., Kumar, A. and Brahmanand, P.S. (2014). Unravelling the spatio-temporal pattern of irrigation development and its impact on Indian agriculture. Irrigation and Drainage, 63: 1-11.

Suresh, A., Aditya, K.S., Jha, G. and Pal, S. (2018). Micro irrigation development in India: an analysis of distributional pattern and potential correlates. International Journal of Water Resources Development, 35 (6):999-1014.

Swaminathan. M.S. (2006). Report of sub-committee on more crop and income per drop of water. Advisory Council on Artificial Recharge of groundwater, Ministry of Water Resources, Government of India.

Theis, S, N Lefore, R Meinzen-Dick and E Bryan (2018). What happens after technologyadoption? Gendered aspects of small-scale irrigation technologies in Ethiopia,Ghana, and Tanzania. Agriculture and Human Values, 35(3), 671–684. https://doi.org/10.1007/s10460-018-9862-8.

Tiwari, K.N., Kumar, M., Santosh, D.T., Singh, V.K., Maji, M.K. and Karan, A.K. (2014). Influence of drip irrigation and plastic mulch on yield of Sapota, (Achraszapota) and soil nutrient. Irrigation and Drainage Systems Engineering, 3 (1):1-8.

Upadhyay, B (2004). Gender aspects of smallholder irrigation technology: Insights fromNepal. Journal of Applied Irrigation Science, 39(2), 315–327.

Van Westarp, S., Chieng, S., Schreier, H., 2004. A comparison between low cost drip irrigation, conventional drip irrigation and hand watering in Nepal. Agric. Water Manage. 64 (2), 143–160.

Vanitha, K. and Mohandass, S. (2014). Effect of humic acid on plant growth characters and grain yield of drip fertigated aerobic rice. The Bioscan, 9:45-50.

Verma, S., Tsephal, S., Jose, T., 2004. Pepsee systems: grass root innovation underground water stress. Water Policy 6, 1–16.

Wrachienb, D., Medicia, M. and Lorenzinia, G. (2014). The great potential of micro irrigation technology for poor-rural communities. Irrigation & Drainage Systems Engineering,3(2):1-2.

2

Organic Saffron Production Possibilities and Approaches

Krishna Kumar[1], Vishal Gupta[1],* Gayatri Jamwal[1], Akash Sharma[2], P.K. Rai[2], V.K.Razdan[1], Satish Kumar Sharma[3], Sushil Kumar Gupta[4] and Shahid Ahmad[5]

[1]*Division of Plant Pathology, Faculty of Agriculture, Sher-e-Kashmir University of Agricultural Sciences & Technology of Jammu Chatha-180009,*
[2]*Advanced Centre for Horticultural Research, Sher-e-Kashmir University of Agricultural Sciences & Technology of Jammu, Udheywalla-180018*
[3]*Regional Horticulture Research Sub-Station, Sher-e-Kashmir University of Agricultural Sciences & Technology of Jammu, Bhaderwah-182222*
[4]*Division of Agroforestry, Faculty of Agriculture, Sher-e-Kashmir University of Agricultural Sciences & Technology of Jammu, Chatha-180009*
[5]*Directorate of Research, Sher-e-Kashmir University of Agricultural Sciences & Technology of Jammu, Chatha-180009*

Globally saffron (*Crocus sativus* L.) is cultivated in regions spanning from the Mediterranean in the west to Kashmir in the east. Countries like Iran, Spain, Italy, India, Greece, Turkey, Afghanistan, Morocco and Azerbaijan harbour suitable climatic and soil conditions required for its growth and harvesting (Mehdi and Reza, 2012). However, the crop is also cultivated on small scales inFrance, Switzerland, Israel, Pakistan, China, Egypt, United Arab Emirates and Japan. Recently its cultivation has been introduced in non-traditional countries like Australia, New Zealand, USA, Argentina and Chile (Fernandez, 2004). Globally, about 85 thousand hectares of land is under saffron cultivation, giving a produce of 288 MT. Iran leads with 90 per cent of total global share for saffron production (Baghalian *et al.,* 2010) and 70% of global export earning a revenue of US$ 51million. India's global share is about 4.5 per cent in production followed by Greece (2.5%), Morocco (0.9%), Spain (0.8%), Italy and China (0.3% each), Afghanistan (0.2%) and Azerbaijan (0.06%).

Apart from being used for religious purposes,the dried stigma of crocus flower, commercially known assaffron, has also been utilized in medicinal and culinary preparations primarily due to its therapeutic and flavour enhancing properties, respectively. Recently, due to indepth scientific exploration of the biological effects of the components of saffron and their potential medical applications,particularly for their therapeutic, cytotoxic, anti carcinogenic and antitumor properties,there has been an increase in demand for saffron consumption(Milajerdi *et al.*, 2016; Razak *et al.*, 2017).

Saffron contains a number of chemical substances exhibiting medicinal and aromatic properties. The quality of saffron depends upon the content of picrocrocin (β-D-glucosida of hydroxysafranal) which imparts a bitter taste, crocin (crocetindigentiobiose ester) which gives bright red colour and safranal which is responsible for aroma. Presence of carbohydrates, minerals, mucilage, vitamins (especially riboflavin and thiamine) and pigments including anthocyanin, carotene, lycopene, flavonoids, amino acids, proteins, starch, gums and other chemical compounds have also been reported in saffron (Abdullah, 1993; Rios *et al.*, 1996; Winter halter and Straubinger, 2000).

In traditional medicines, saffron has been used as a stomach pain soother, antispasmodic, digestion aid, renal colic pains reliever, and anti-depressant (Guo *et al.*, 2022). Saffron is rich in riboflavin (56-138 μgg^{-1}) and thiamine (0.7–4 μgg^{-1}) and also have considerable number of lipophilic carotenoids. Saffron acts against cancer formation and exhibit selective toxicity against tumours by directly targeting DNA and regulating gene expression and induction of cell apoptosis (Abdullaev, 2002; Milajerdi *et al.*, 2016). Saffron affects neurotransmitters level such as norepinephrine, dopamine and serotonin by decreasing the activity of acetylcholine esterase and inhibiting the activity of monoamine oxidase in the brain (Jelodar *et al.,* 2018). Presence of β-carotene in saffron has effect on the structure and function of photoreceptors which improve the visual function of ARMD (Riazi *et al.,* 2017).

The Union Territory of Jammu and Kashmir enjoys the monopoly of growing saffron crop in India, althoughit forms only 7 per cent of the total global area under saffron cultivation. About 16,269 families are associated with the cultivation of this high value but low volume crop in India (Dass and Deshpande, 2018). In the J&K, its cultivation is confined to the districts of Pulwama, Budgam and Srinagar of Kashmir province and district Kishtwar of Jammu province. Pulwama, Budgam and Srinagar have 75.00, 16.13 and 6.68 per cent of total areaunder saffron cultivation respectively, whereas, Kishtwar has only 2.5 per cent (Nehvi, 2009). Areas in Pulwama, Budgam, Srinagar and Kishtwar districts, account for 56.83, 30.62, 11.36 and 1.20 per cent of total production, respectively. Efforts also have been made to introduce saffron cultivation in hilly regions of Uttarakhand (Uttranchal), Himachal Pradesh and Sikkim (Nauriyal *et al.,* 1977; Koul, 1999;

Rana *et al.,* 2003). Export of saffron plays an important role in the economy of J&K UT. The 2012, Saffron export worth about $2 million was recorded in India. In 2020, 1.4 million dollars of saffron was exported to UAE, US, Australia, Vietnam and Nepal, making India 10th largest exporter of the world (oec.world).

Saffron cultivation in J&K is under threat of annihilation which is apparent from its dwindling global production share. During last few years due to various abiotic and biotic stresses, industrialization, urbanization, lack of mechanization, unorganized market and fluctuating market prices,area,production and productivity of saffronhas declined from 5707 to 3715 ha, 16 to 9.5 MT and 3.13 to 2.00 kg/ha, respectively (Anonymous, 2014). In 2020 we have imported 23.1 million dollar worth of saffron, ranking fourth in the list of worlds largest importers of saffron (oec.world).

The current productivity level of saffron in the district of Kishtwar is 1.5-2.0 kg/ha, which is much lower when compared to the productivity levels of other countries of the world (7.5 kg/ha). Further, during the last two years the saffron production has shown a steep decline. Among various factors responsible for low productivity, corm rot complex is one of the major constraints in successful cultivation of saffron. The incidence of corm rotranges from 15 to 100 per cent, which has been aggravating year after year (Gupta and Vakhlu, 2015). Multiple soil-borne pathogens like *Fusarium oxysporum* f. sp. *gladioli, F. oxysporum, F. solani, F. equiseti, F. monoliforme*var. *intermedium, Rhizoctoniasolani, Rhizoctoniacrocorum, Phomacrocophila, Macrophominaphaseolina, Mucor* sp., *Penicillium* sp. and *Sclerotium rolfsii* have been reported to be associated with the disease (Nehvi *et al*., 2006; Husaini *et al.,* 2010; Gupta *et al.,* 2020).

Corm rot disease can be managed by the use of chemical fungicides *viz*., carbendazim, mycobutanil, mencozeb and Tecto (Husaini *et al*., 2010). However, their continuous application not only has deleterious impact on environment, beneficial micro-flora and human health (Aktar *et al*., 2009), but it is also responsible for the evolution of resistant pathotypes of the pathogens (Hobbelen *et al*., 2014). Application of biological control methods has emerged as a sustainable alternative for strategies employed in managing plant pathogens. It reduces the disease producing activity of pathogens through parasitism, antibiosis, competition, lytic enzymes and induced resistance (Cook and Baker, 1983; Pal and Gardener, 2006), besides, having no deleterious impact on human health or environment (Baker and Paulitz, 1996). Biocontrol may be achieved by introducing a prospective microorganism or using indigenous microorganisms that have potential to delimit the survival of the pathogen (Gupta and Vakhlu, 2015; Gupta *et al.,* 2021). Bacteria and fungi isolated from rhizosphere and bulk soil communities have been reported to be effective biocontrol agents (Whipps *et al*., 2001). Further, soil solarisation is a soil disinfestation method for managing soil borne plant pathogens. It does not release any dangerous chemical residue and is considered as a safe, simple, effective and eco-friendly tool for nurseries and small holdings (Kapoor, 2013; Gill *et al.,* 2017). Unique features of the corm rot disease, particularly the persistence of inoculums in the soil, has led to devising systems having holistic approach to manage the disease, which can be viable in terms of technological, economic, social and regulatory aspects, as well as knowledge transferable.

Status of saffron corm rot

Corm rot was first time reported from Japan (Yamamoto *et al.,* 1954), and then in India (Shah and Srivastava, 1984), Spain (Garcia-Jimenez and Alfaro Garcia, 1987) and Italy (Cappelli, 1994). Currently, the disease is widespread, causing substantial yield losses in several saffron-growing regions of the world (Primo and Cappelli, 2000; Ayatollah, 2014; Ambardar *et al*., 2016; Bushra and Shoib, 2018; Najari *et al*., 2018). Rekhi *et al*. (1990) studied the corm rot of saffron in Kashmir division and recorded that the disease was main constraint among the farmers for successful cultivation of saffron. Corm rot incidence was also recorded in Himachal Pradesh (Sud *et al*., 1990). Shah and Srivastava (1984), during a survey in Almora (Uttarakhand)found rotted corms associated with fungus. In Abruzzi, Italy, the disease was present in 54 per cent surveyed saffron fields, causing 50 per cent plant mortality during severe outbreaks of the disease. About 40 per cent disease severity and 100 per cent diseases incidence of corm rot of saffron was observed in Kashmir, causing reduced plant growth and yield (Gupta and Vakhlu, 2015).

The corm rot of saffron is characterized by symptoms such as yellowing and wilting of leaves and shoots due to basal stem rot, development of black round spots on the corm, drooping of leaves in infected corms and in severe cases conversion of whole corm into black powdery mass causing it to decompose (Palmero *et al*., 2014). Symptoms of the disease are mainly observed during the flowering stage (October-November). Roots attached to infected corms also become brown and decay. Severely infected corms rot before they sprout. Plants arising from less severely infected corms exhibit "yellows" symptoms late in the season. Older foliage gradually turns yellow from the tip towards back and eventually all foliage turns yellow and die (Gupta *et al.,* 2011). In newly infested fields, the disease occurs in small patches that gradually enlarges with each year till the whole field is diseased.

Multiple necrotrophic and hemiobiotrophic pathogens have been associated with corm rot complex, which included *Sclerotium rolfsii* (Kalha *et al*., 2007), *Macrophomoinaphaseolina* (Carta *et al*., 1982; Hassan and Devi, 2003; Thakur *et al*., 1992; Ionita *et al.,*1995), *Fusarium oxysporum* f. sp. *gladioli* (Yamamoto *et al.,* 1956; Shah and Srivastava, 1984; Garcia-Jimenez *et al.,* 1985; Garcia-Jimenez and Alfaro-Garcia, 1987; Cappelli, 1994), *Fusarium solani* (Sud *et al*., 1999), *Pencilliumcorymbiferum* (Gu and Zhi, 1997; Capelli, *et al*., 1991; Ahmad and Sagar,

2007), *Phomacrocophila*. (Nannizzi, 1941), *Rhizoctoniaviolacea* var. *crocorum* (Nannizzi, 1941), Bacillus*croci* (Mizusawa, 1923) and *Burkholderiangladioli* (Xu and Ge, 1990; Fiori *et al.*, 2011), *Porostereum* sp., (Wani *et al.*, 2018) and a new *formaespeciales* saffrani (*Palmero et al. (2014)*

Fusarium oxysporum f. sp. *gladioli* is capable of surviving in infected corms and soil as mycelium, chlamydospore, microconidia and macroconidia (Brayford, 1996) and is disseminated to un-infested saffron growing areas through diseased corms (Cappelli and Di Minco, 1999). Due to the vegetative mode of propagation of saffron, sequential infection initiates through the corm by germinating spores or by mycelium which gain entry directly through roots or through wounds caused by insects, nematodes, rodents or mechanical abrasion. Later, resting structures of the pathogens are formed in the infected host tissues. When crop residues containing the resting structures are incorporated into saffron field soils, intensive microbial activity occurs and successions of microorganisms develop in the decomposing plant tissues (Katan, 2017). Survival of soil-borne pathogens responsible for corm rot can be active or passive for various years in the absence of host. Further, the pathogens also possess the autonomous and passive mode of dispersal which is helpful in the spatial-temporal distribution of corm rot disease in saffron.

Lack of irrigation facilities, lack of resistant source, frequent drought, poor drainage, mono-culturing of saffron, coupled with continuous cultivation of same germplasm has resulted in the frequent occurrence of corm rot and mass mortality of the plants. Further, the pathogen inoculum is introduced into new saffron-growing regions through contaminated corms (Cappelli and Di Minco, 1999). Corm rot adversely affects saffron production and simultaneously impedes the production of daughter corms (Kalha *et al.*, 2008).

Biological control and plant growth promotion in saffron

1. Use of Plant growth promoting rhizobacteria and fungi

Biological control isrecognised as an effective, eco-friendly, sustainable and alternative approach for disease management of crops. Currently plant growth promoting rhizobacteria (PGPR) and plant growth promoting fungi (PGPF) are being exploited commercially for their bio-protectant, bio-fertilizers, bio-remediation and bio-stimulant potential. Sharaf-Eldin *et al.* (2008) found that soaking of corms and drenching of soil with *Bacillussubtilis* (FZB24) significantly increased the leaf length, weight of stigma and also reduced the number of days to sprout as compared to control along with enhancement in picrocrocin, crocetin and safranalcontent. In field experiments, the soil application of talc-based formulation of *Trichoderma viride* and corm treatment with carbendazim were found to be effective in reducing the disease incidence of saffron and simultaneously increasing germination of corms(Gupta *et al.*, 2011). Mir *et al.*

(2011) recorded thatseveral native isolates of *Trichoderma* spp. significantly inhibited and overgrew the mycelial growth of the different pathogensassociated with corm rot such as *Fusariumoxysporum* f. sp. *gladioli,* Basidiomycetes fungus, *Fusarium solani*, *Rhizoctoniasolani*, *Phytophthora* spp. and *Fusariumoxysporum*.

Fifty-four bacterial endophytes isolated from saffron plant were identified using molecular approaches and grouped into eleven different taxa out of which *B. licheniformis* was found as the dominant endophyte present in both leaves and corms of saffron. Eighty-one per cent of these isolates showed lipase activity, 57 per cent cellulase, 48 per cent protease, 38 per cent amylase, 33 per cent chitinase and 29 per cent pectinase activity. Further, out of total, 24 per cent exhibited phosphate solubilising ability, 86 per cent showed siderophore production and 80 per cent have phytohormone potential (Sharma *et al*. (2015).

Gupta and Vakhlu (2015) isolated *Bacillus amyloliquefaciens* strain W2 from saffron field and found it possessed antagonistic activity against *Fusarium oxysporium*, *F. solani*and *Penicillium* sp. by inhibition of mycelial growth. Nehvi*et al*. (2010)studied the impact of microbial inoculation on morphological and flowering tributes of saffron crop and found that the application of bio-fertilizer (*Azospirillum* and VAM) increased the leaf area and plant height which may be due to the production of growth promoting substances and fixation of nitrogen.

Wani *et al*. (2017) isolated beneficial fungal endophyte, *Mortierellaalpina* CS10E4, from saffron cropduring dormant and vegetative stages, which possessed the biocontrol activity and plant growth promotion traits. *M. alpina* CS10E4 showed growth inhibition of *F. oxysporum*, significant increase in total biomass, number of apical sprouting buds, number of adventitious roots and two-fold increase in chlorophyll content as compared to un-inoculated treatments. In flower tissues, phenolic and flavonoid content increased and the crocin and safranal content were further increased by two and four-fold in *M.alpina* CS10E4 treated corms.

Al-Ahmadi *et al*. (2018) isolated plant growth rhizobacteria from the rhizosphere of the saffron fields, which were identified on the basis of phenotypic traits and 16S rDNA sequences analysis. Eighteen bacterial isolates were identified representing twelve species, which under field conditions resulted insignificant increase in leaf area especially with the inoculation of *P. fluorescens*, *Paenibacillus* spp. (BA5) and *Pectobacterium* spp.

Arbuscular Mycorrhizal Fungi (AMF) constitute another group of root obligate biotrophs known to protect the host plant from several soil-borne diseases (Azcon-Aguilar *et al*., 2002).They employ different mechanisms likebiochemical changes in plant tissues, microbial changes in rhizosphere, nutrient status, anatomical changes to cells, changes to root system morphology and stress alleviation for biocontrol (Hooker *et al*., 1994; Berruti *et al*., 2016). Shuab *et al*., (2016) recorded substantial increase in root colonisation by AMF (*Glomusintraradices* and *Glomusmosseae)* structures at 30 days of sowing of saffron. Population of AMF gradually increased with the age of plant, maximising into root colonisation. Scales of saffron corms of inoculated plants showed AMF colonisation after 60 days of bud emergence. The height, number of leaves per plant, leaf length, cormlet number, fresh and dry weight of root and shoot significantly increased in AMF inoculated plants.

One of the prerequisites of plant growth promoting bacteria for biocontrol is their successful establishment and persistence in the rhizosphere or within plant root tissues. Root colonization, rhizosphere competence or niche not only results in high population densities of the beneficial microorganisms on the root system, but it also functions as the delivery system of antifungal metabolites (AFMs) (Lugtenberg *et al.*, 1999; Landa *et al.*, 2003). Siderophoreproducing plant growth promoting rhizobacteria improve plant health and reduce plant diseases by improving iron nutrition for the host, inhibiting the growth of other micro-organisms with the release of their antibiotic moleculeshinder the growth of pathogens by limiting the iron availability for the pathogen (Goswami *et al.,* 2016).

Production of antibiotic is the most important mechanism employed by beneficial microbes for biocontrol. For example, for the suppression of phytopathogens, different strains of *Pseudomonas* and *Bacillus* spp. produce lytic enzymes which are able to hydrolyze chitin, proteins, cellulose and hemicellulose present in the cell wall of fungi and oomycetes (Nielsen and Sorensen, 1999; Lim *et al.,* 1991; Kumar *et al.,* 2010). Some *Pseudomonas* sp. can enhance the photosynthesis activity through producing phytohormones, thus increasing phosphorus absorption by plant, nitrogen fixation and synthesis of enzymes which modifying ethylene level in plants (Mayak *et al.,* 2004).

An alternate strategy possessed by biocontrol agents to protect the crop plant isby increasing the level of defence related transcripts. This is called as induced systemic resistance (ISR) and systemic acquired resistance (SAR). It has been studied in plants like *Arabidopsis* (van de Mortel *et al.,* 2012), tobacco (Maurhofer *et al.,*1994), rice (Pieterse *et al.,* 2000) and wheat (Maketon *et al.,* 2012). Mutual recognition and substantial coordination of plant and microbial responses are important for the establishment of beneficial associations which subsequently modulate plant immunity (Mikicinski *et al.,* 2016). Phosphate solubilisation, siderophore production, nitrogen fixation and degrading starch to glucose are among the mechanisms used by which *B. subtilis* toimprove plant growth (Huang *et al.*, 1998; Rai, 2006; Lugtenberg and Kamilova, 2009). The *Trichoderma* spp. *uses m*ycoparasitism to parasitize and kill plant pathogenic fungi(Harman *et al.,*2004), colonize plant roots, both externally and internally (Bailey *et al.,* 2006), produce volatiles, enzymes and toxins ((Padmodaya and Reddy, 1996; Dubey and Patel, 2001) and induces of systemic resistance in host (Djonovic *et al.,* 2007). That is why they are commercially successful as biofungicides.

2. Soil solarization

Growing concern about environmental hazardous and human health issues by pesticide usedirected to develop environmentally friendly management strategies against plant diseases.Soil solarisation was first introduced in 1976 (Katan *et al*., 1976; Gamliel and Katan, 2012a) to control soil-borne pathogens and has emerged as an alternative for chemical pest management. It is a chemical-free way of managing pest population in the soil before planting (Katan, 1987; McGovern and McSorley, 1997; Gill *et al.*, 2009) along with improving soil texture and the availability of nutrientsby trapping the solar energy, which raises the temperature. Soil solarization successfully manages the soil-borne fungal and bacterial plant pathogens in different cropping systems, such as *Rhizoctonia* spp., *Fusarium* spp., *Sclerotinia* spp., *Macrophomina* spp., *Phytophthora* spp., *Verticillium* spp., *Agrobacterium tumefaciens, Clavibactermichiganensis, Pythium* spp., and *Streptomyces* scabies (McGovern and McSorley, 1997; Chellemi and Mirusso, 2006; Gelsomino *et al*., 2006).

Barakat and Al-Masri (2012) during their study recorded that regular polyethylene (PE) sheets, double polyethylene (DPE) sheets and virtually impermeable films (VIF) increased the soil temperatures. The solarization by PE, DPE and VIF slightly increased the pH, EC, available calcium and potassium compared to the control. Double polyethylene (DPE) sheets reduced disease incidence of Fusarium wilt of tomato and the population of *Fusarium oxysporum*f. sp. *Lycopersici*thereby increasing fresh and dry plant weight. The reduction in pathogen population may be due to the increase in temperature, killing the pathogen directly or due to weakening the pathogen structures with sub-lethal heat, rendering it incapable of inducing crop damage (DeVay and Katan, 1991). Concentrations of NH_4, NO_3andother soluble mineral nutrients, including calcium, magnesium, phosphorus, potassium and othersin thetop 15 cm of soil also increasesitssolarization, which ultimately increases growth response (IGR). This can be attributed to a number ofmechanisms, including the increase in nutrient levels in soilsolution, stimulation of beneficial microorganisms and thecontrol of minor pathogens (Stapleton and DeVay, 1995).

Conclusion

The corm rot not only reduces yield but it also affects the quality and quantity of daughter corms. Intensive cultivation, poor agronomic practices, mono-culture, drought, lack of resistance sources along with lack of breeding approaches, longer planting cycle and continual use of diseased propagating material in saffron growing areas has led to increased incidences and severity of corm rot disease. This has subsequently reduced the production of healthy daughter corms, thereby limiting the availability of propagating material for the cultivation in the next season (Husaini *et al*., 2010).

Considering the prevalence, ubiquitous nature and the ability of soil-borne fungi to cause epidemics in relatively short period of time, disease management strategies are needed to secure the productivity in today's agriculture. Several beneficial soil organisms have been identified with promote plant growth and act as defensive agents against pathogens. A detailed understanding of their working mechanism with respect to their interaction with both plants and pathogens would help us in devising effective methods of disease control in plants. Solarization is another useful approach for disease reduction and plant growth improvement, which will definitely open doors for a wide use of the technique in regions characterizedwith lower average temperatures and on small holdings. The use of fungicides also plays a major role in increasing the productivity of several crops including saffron, throughout the world by suppressing the plant diseases. Therefore, a combinatorial approach towards pathogen control would help in effective disease management in plants thereby increasing growth and productivity

References

Abdullaev, F.I. 2002. Cancer chemopreventive and tumoricidal properties of saffron (*Crocus sativus* L.). Experimental Biology and Medicine, 227: 20–25.

Abdull*ah, F.I. 1993.* Biological effects of saffron. Biofactors,4: 83-86.

Ahmad, M., Bashir, S., Wani, A. and Sagar, V. 2008. Corm Rot of Saffron (*Crocus sativus* L) and its management. In: Nehvi, F. A. and Wani, S. A. (eds.). Saffron Production in Jammu and Kashmir.pp 232-241. Universal Printers Barzulla, Srinagar.

Ahmad M and Sagar V.2007. *Annual Progres*sReport. Integrated management of corm/ tuber rotof Saffron and Kalazera. Horticulture TechnologyMini mission-I Research Project No.2.2. pp 22. ICAR, India

Aktar, W., Sengupta, D. and Chowdhury, A. 2009. Impact of pesticides use in agriculture: their benefits and hazards. Interdisciplinary Toxicology, 2(1):1-12.

Ambardar, S. Singh, H.R., Gowda, M. and Vakhlu, J. 2016. Comparative metagenomics reveal phylum level temporal and spatial changes in mycobiome of belowground parts of Crocus sativus. PlosOne, 11(9):0163300.

Anonymous, 2014. Saffron Industry Value ChainDevelopment. In IranDiagnostic Study Report United Nations Industrial Development OrganizationVienna International Centre, P.O. Box 300, 1400 Vienna, Austria. pp 1-80.

Ayatollah, S. 2014. Identification of some saffron corm rot fungi and their control. Saffron Agronomy and Technology, 2:205-213.

Azcon-Aguilar, C., Jaizme-Vega, M.C. and Calvet, C. 2002. The contribution of arbuscular mycorrhizal fungi to the control of soil borne plant pathogens. In: Gianinazzi, S. and Schuepp, H. (eds.). Mycorrhizal Technology: From Genes To Bioproducts-Achievement and Hurdles in Arbuscular Mycorrhizal Research. pp 187-198. Brikhauser, Basel, Switzerland.

Baghalian, K., Shabani, S. M. and Jamshidi, A.H. 2010. Genetic variation and heritability of agro-morphological and phytochemical traits in Iranian saffron (*Crocus sativus* L.) populations. Industrial crops and product ,31:401–406.

Bailey, B. A., Bae, H., Strem, M.D., Roberts, D.P., Thomas, S. E., Crozier, J., Samuels, G. J., Choi, I. Y. and Holmes, K. A. 2006. Fungal and plant gene expression during the colonization of cacao seedlings by endophytic isolates of four *Trichoderma* spp. Planta, 224:1449–1464.

Baker, R. and Paulitz, T.C. 1996. Theoretical basis for microbial interactions leading to biological control of soil-borne plant pathogens In: Hall, R. (ed). Principles and Practice of Managing Soil-borne Plant Pathogens. pp50-79. American Phytopathological Society, St. Paul, Minnisota.

Barakat, R, M. and Al-Masari, M. I. 2012. Enhanced soil solarization against *Fusarium oxysporum* f. sp.lycopersici in the uplands. International Journal of Agronomy. 1-7.

Berruti, A., Lumini, E., Balestrini, R. and Bianciotto, V. 2016. Arbuscular mycorrhizal fungi as natural biofertilizers: let's benefit from past successes. Frontiers in Microbiology,6: 1559.

Booth, C (ed.). 1971. The genus *Fusarium*. pp 237. Common wealth Mycological Institute, Kew, England.

Brayford, D. 1996. *Fusarium oxysporum* f. sp. gladioli. Mycopathologia, 133: 47-48.

Bushra, J and Shoib, A. B. 2018. Corm Rot of Saffron: Symptoms and Biological management. EC Microbiology, 14: 01.

Cappelli, C., and Di-Minco, G. 19*99. Results of a t*riennial study on saffron diseases in Abruzzi. Informatore Fitopatologico,49: 27-32.

Cappelli, C. 1994. Occurrence of *Fusarium oxysporum* f. sp. gladioli on saffron in Italy. Phytopathologia Mediterranea, 33: 93–94

Carta, C., Fiori, M. and Franceschini, A. 1982. Il "marciume carbonioso" dei bulbi dello zafferano (*Crocus sativus* L.). Studi Sassaresi, Annali della Facoltà di Agraria, Università di Sassari XXIX, 193–197.

Chellemi, D. O. and Mirusso, J. 2006. Optimizing soil disinfection producers for fresh market tomato and pepper production. Plant Diseases, 90, 668–674.

Cook, R. J. and Baker, K. F. 1983. The Nature and Practice of Biological Control of Plant Pathogens, pp 539. American Phytopathological Society, St. Paul, Minnisota.

Dass, A. A. and Deshpande S. V. 2018. Saffron cultivation: area, production and productivity in district Pulwama (J&K).New Man International Journal of Multidisciplinary Studies, 5: 45-48.

De Vay, J. E. and Katan, J. 1991. Mechanisms of pathogen control in solarized soils. In: Katan, J. and De Vay, J. E.(eds.) Soil Solarization. pp 87-102. CRC Press, Boca Raton, Florida, USA.

Djonovic, S., Vargas, W.A., Kolomiets, M. V., Horndeski, M., Wiest, A., Kenerley, C. M., 2007. A proteinaceous elicitor Sm1 from the beneficial fungus Trichoderma virens is required for induced systemic resistance in maize. Plant Physiology. 145:875–889.

Fernandez, J.A. 2004. Biology, biotechnology and biomedicine of saffron. Recent Research Development in Plant Sciences. 2: 127–159.

Fiori, M., Ligios,V. and Schiaffino A.2011. Identification and characterization of Burkholderia isolates obtained from bacterial rot of saffron (*Crocus sativus* L.) grown in Italy. Phytopathology Mediterranian,50: 450-461.

Gamliel, A. and Katan J., 2012a. Soil Solarization: Theory and Practice. APS Press, USA.

Garcia-Jimenez, J., Piera, V.J. and Alfaro-Garcia, A. 1985. Los "soldatos", una nueva enfermedad del azafaran (*Crocus sativus* L.) en Espana. Proceed[in]gs IV Congreso Nacional de Fitopatologia. Sociedad Espanola de Fitopatologia. Pamplona, Spain, 76.

García-Jiménez, J. and Alfaro-García, A. 1987. Fusarium oxysporumSchlecht. as causal agent of a seed-borne disease of saffron (*Crocus sativus* L.). 7th Congress of the Mediterranean Phytopathological Union. pp 15. Granada, Spain,

Gelsomino, A., Badalucco, L., Landi, L. and Cacco, G. 2006. Soil carbon, nitrogen and phosphorus dynamics as affected by solarization alone and combined with organic amendment. Plant Soil, 279: 307–325.

Gill, H. K., Aujla, I. S., De Bellis, L. and Luvisi, A. 2017. The role of soil solarization in india: how an unnoticed practice could support pest control. Frontier Plant Sciences. 8:1515.

Gill, H. K., McSorley, R. and Treadwell, D. D. 2009. Comparative performance of different plastic films for soil solarization and weed suppression. Horticulture Technology,19: 769-774.

Goswami, D., Thakker, J.N., Dhandhukia, P.C. and Moral, M.T.2016. Portrayingmechanics of plant growth promoting rhizobacteria (PGPR): A review. Cogent Food & Agriculture,2:1-19.

Gu, Z.F. and Zhi, Y. 1997. A study on biological characteristics of Penicillium corymbiferum on Crocus. Acta Agriculturae Shanghai, 13: 71-74.

Guo Z.L., Li M.X., Li X.L., Wang P., Wang W.G., Du W.Z., Yang Z.Q., Chen S.F., Wu D. and Tian X.Y. 2022. Crocetin: A Systematic Review. Front. Pharmacol. 12:745683.

Gupta, R. and Vakhlu, J. 2015. Native Bacillus amyloliquefaciens W2 as a potential bio-control for Fusarium oxysporum R1 causing corm rot of Crocus sativus. European Journal of Plant Pathology, 143:123-131

Gupta, V., Kumar, K., Fatima, K., Razdan, V.K., Sharma, B.C., Mahajan, V., Rai, P.K., Sharma, A., Gupta, V., Hassan, M.G. and Hussain, R. (2020). Role of biocontrol agents in management of corm rot of saffron caused by Fusarium oxysporum. Agronomy. 1398. DOI: 10.3390/agronomy10091398.

Gupta, V., Sharma, A., Rai, P.K., Gupta, S.K., Singh, B., Sharma, S.K., Singh, S.K., Hussain, R., Razdan, V.K., Kumar, D., Paswal, S., Pandit, V. and Sharma, R. (2021). Corm Rot of Saffron: Epidemiology and Management. Agronomy. 11, 339. DOI:10.3390/agronomy11020339.

Gupta. V., Kalha, C.S., Razdan, V.K. and Dolly. 2011. Etiology and management of corm rot of saffron in Kishtwar district of Jammu and Kashmir, India. Journal of Mycology & Plant Pathology, 41(3):361-366.

Harman G.E., Howell C.R., Viterbo A., Chet I. and Lorito, M. 2004. Trichoderma species-opportunistic, avirulent plant symbionts. Nature Review Microbiology, 2: 43–56.

Hassan, M. G. and Devi, L. S. 2003. Corm rot diseases of saffron in Kashmir valley. Indian Phytopathological, 56: 122.

Hobbelen, P.H.F., Paveley, N.D. and Bosch, F.2014. The emergence of resistance to fungicides. PlosOne. 3: 9.DOI: 10.1371/journal. pone.0091910

Hooker, J. E., Jaizme-Vega, M. and Atkinson, D. 1994. Bio-control of plant pathogens using arbuscular mycorrhizal fungi. In: Gianinazzi, S. and Schüepp, H. (eds.). Impact of arbuscularmycorrhizas on sustainable agriculture and natural ecosystems. pp 191–200. Brikhauser, Basel, Switzerland.

Huang, J. W., Blaylock, M. J., Kapulnik, Y. and Ensley, B. D. 1998. Phytoremediation of uranium-contaminated soils: role of organic acids in triggering uranium hyperaccumulation in plants. Environmental Science and Technology,32: 2004-2008.

Husaini, A. M., Hassan, B. and Ghani, M. Y. 2010. Saffron (Crocus sativus L. Kashmirianus) cultivation in Kashmir: practices and problems. Functional Plant Sciences Biotechnology, 4: 108–115.

Ionita A., Iliescu, H., Jinga, V. and Iordache, E. 1995. Macrophominaphaseolina a dangerous parasite of cropped plants possibilities for control. Probleme de Protectio Plantelor, 23: 179-196.

Jelodar, G. A., Javid, Z. Sahraian, A.andJelodar, Sina.2018. Saffron improved depression and reduced homocysteine level in patients with major depression: A Randomized, double-blind study. Avicenna Journal of Phytomedicine,8: 43–50.

Kalha, C. S., Gupta, V., Gupta, D. and Mohiddin, F.A. 2008. An overview of saffron cultivation with special reference to district Kishtwar. In: Nahvi, F.A. and Wani, S.A (eds.). Saffron Production in Jammu and Kashmir Directorate of Extension Education. pp 142-152. SKUAST-K, Srinagar.

Kalha, C. S., Gupta, V., Gupta, D. And S. 2007. First report of sclerotial rot of saffron caused by Sclerotium rolfsii in India. Plant Disease, 91: 1203-1206.

Kapoor R. T. 2013. Soil solarization: eco-friendly technology for farmers in agriculture for pest management. Proceedings of 2nd International Conference on Advances in Biological and Pharmaceutical Sciences (ICABPS 2013). Hong Kong.

Katan, J. (ed.) 1987. Soil solarization, In: Innovative Approaches to Plant Disease Control. pp 77-105. John Wiley and Sons, New York.

Katan, J., Greenberger, A., Alon, H. and Ginstein, A. 1976. Solar heating by polyethylene mulching for the control of diseases caused by soil-borne pathogens. Phytopathology, 66: 683-689.

Katan, J. 2017. Diseases caused by soil borne pathogens: biology managements and challenges. Journal of Plant Pathology.99: 305-315.

Koul, A. K. 1999. Saffron present status and propagation. Proceedings of National symposium onSaffron, pp 21–23. Regional Research Laboratory, Jammu, India.

Kumar, H. Bajpai, V. K. and Dubey, R. C. 2010. Wilt disease management and enhancement of growth and yield of Cajanuscajan (L) var. Manak by bacterial combinations amended with chemical fertilizer. Crop Protection, 29: 591-598.

Landa, B. B., Mavrodi, D. M., Thomashow, L. S. and Weller, D. M. 2003. Interactions between strains of 2, 4-diacetylphloroglucinol producing Pseudomonas fluorescens in the rhizosphere of wheat. Phytopathlogy, 93: 982-994.

Lim, H. S., Kim, Y. S. and Kim, S. D. 1991. Pseudomonas stutzeri YPL-1 genetic transformation and antifungal mechanism against Fusarium solani, an agent of plant root rot. Applied and Environmental Microbiology, 57: 510-516.

Lugtenberg, B. and Kamilova, F. 2009. Plant-growth-promoting rhizobacteria. Annual Review of Microbiology, 63:541–556.

Lugtenberg, B.J., Kravchenko, L. V. and Simons, M. 1999. Tomato seed and root exudates sugars: composition, utilization by Pseudomonas biocontrol strains and role in rhizosphere colonization. Environmental Microbiology, 1: 439-446.

Maketon, C., Fortuna, A. M. and Okubara, P. A. 2012. Cultivar dependent transcript accumulation in wheat roots colonized by Pseudomonas fluorescens Q8r1-96 wild type and mutant strains. Bio-Control, 60: 216-224.

Maurhofer, M., Hase, C., Meuwly, P., Metraux, J. P. and Defago, G. 1994. Induction of systemic resistance of tobacco-to-tobacco necrosis virus by the root-colonizing Pseudomonas fluorescens strain CHA0: influence of the gacA gene and of pyoverdine production. Phytopathology,84:139-146.

Mayak, S., Tirosh, T. and Glick, B.R. 2004. Plant growth promoting bacteria that confer resistance in tomato to salt stress. Plant Physiology and Biochemistry, 42, 565-572.

McGovern, R. J. and McSorley, R. 1997. Physical methods of soil sterilization for disease management including soil solarization. In:Rechcigl, N. A. and Rechcigl, J. E. (eds). Environmentally safe approaches to crop disease control. pp. 283–313. CRC Lewis Publishers, Boca Raton, Florida.

Mehdi, S. and Reza, M. 2012. Saffron Export and agricultural value added in Iran, European Journal of Experimental Biology, 2: 24-31

Mikicinski, A., Sobiczewski, P., Puławska, J. and Malusa, E. 2016. Antagonistic potential of Pseudomonas graminis 49M against Erwinia amylovora, the causal agent of fire blight. Archives of Microbiology. 198:531–539.

Milajerdi, A.R., Djafarian, K. and Hosseini, B. 2016. The toxicity of saffron (*Crocus sativus* L.) and its constituents against normal and cancer cells.Journal of Nutrition & Intermediary Metabolism. 3: 23-32.

Mir, G. H., Devi, L. S., Ahmad, S., Kumar, V. M. and Williams, P. 2011. Antagonistic potential of native isolates of Trichoderma viride on corm rot pathogen complex of saffron (Crocus sativus) in Kashmir. Plant Pathology, 10:73-78.

Mizusawa, Y., 1923. A bacterial rot disease of saffron. Annals of the Phytopathological Society of Japan, 1: 1-12

Nannizzi A., 1941. I parassitidellepianteofficinali. Roma, Istituto Poligrafico dello Stato. 1–544

Nauriyal, J.P.,Gupta, R. and George, C.K. 1977. Saffron in India. Arecanut and Spice Bulletin. 8: 59–72.

Nehvi, F.A, Wani, S. A., Dar, S.A., Allaie, B.A. and Mir, Z.A. 2006. Status and opportunities for harnessing high Saffron Productivity in Kashmir. 2nd International Symposium on Saffron, pp 36-37. (Abstracts).

Nehvi, F.A., Lone, A.A., Khan, M.A. and Maqhdoomi, M.I. 2010. Comparative study on effect of nutrient management on growth and yield of saffron under temperate conditions of Kashmir. Acta Horticulture, 850: 165-170.

Nehvi, F.A. 2009. A value chain on Kashmir saffron. Final Report, Project Implementation Unit, NAIP, ICAR, New Delhi.

Nielsen, M. N. and Sorensen, J. 1999. Chitinolytic activity of Pseudomonas fluorescens isolates from barley and sugar beet rhizosphere. FEMS Microbiology Ecology, 30: 217-227.

Padmodaya, B. and Reddy, H.R. 1996. Screening of Trichoderma spp. against Fusarium oxysporumf. sp. lycopersicicausing wilt in tomato. Indian Journal Mycology Plant Pathology, 26: 266–270.

Pal, K. K. and Gardener, B. M. 2006. Biological control of plant pathogens. Plant Health Instructor, 2: 1117–1142.

Palmero, D., Rubio-Moraga, A., Galvez-Paron, L., Nogueras, J., Abato. C., Gomez-Gomez, L. and Ahrazem, O. 2014. Pathogenicity and genetic diversity of Fusarium oxysporum isolates from corms of Crocus sativus. Industrial Crops Products,61: 186–192.

Pieterse, C. M. J., Van Pelt, J. A., Ton, J., Parchmann, S., Mueller, M. J., Buchala, A. J., Metraux, J. P. and Van Loon, L. C. 2000. Rhizobacteria-mediated induced systemic resistance (ISR) in Arabidopsis requires sensitivity to jasmonate and ethylene but is not accompanied by an increase in their production. Physiological and Molecular Plant Pathology. 57:123-34.

Primo, P. Di. and Cappelli,C.2000. Preliminary characterization of Fusarium oxysporumf. sp. gladioli causing Fusarium corm rot of saffron in Italy. Plant Diseases, 84:806.

Rai, M. K. 2006. Handbook of Microbial Biofertilizers. Pp 579.CRC Press. Boca Raton, Florida, USA.

Rana, S. S., Sood, P., Mondal, K. K. and Thakur, K.S., (eds.) 2003. Technology for saffron cultivation. pp. 20.Mountain Agricultural Research and Extension Centre publication: Sangla (Kinnaur), HP, India.

Razak, S. I. A., Hamzah, M.S.A., Yee, C. F., Kadir, M.R.A. and Nayan, N. H.M. 2017. A review on medicinal properties of saffron toward major diseases. Journal of Herbs, Spices & Medicinal Plants. 23:98-116.

Rekhi, S. Kumar, K. and Sagwal, R. C.1990. Research gaps in scientific cultivation of saffron in Kashmir. Agriculture Science Digest. 17:161-164

Riazi, A. Panahi, Y., Alishiri, A. A.,Hosseini, M. A. Karimi Zarchi, A. A. and Sahebkar, A. H. 2017. The impact of saffron (*Crocus sativus*) supplementation on visual function in patients with dry age-related macular degeneration. Italian Journal of Medicine,11: 2

Rios, J. L., Recio, M.C., Giner, R.M. and Manez, S. 1996. An update review of saffron and its active constituents. Phytotherapy Research, 10: 189-193.

Shah, A and Srivastava, K. K. 1984 Control of corm rot of saffron. Progressive Horticulture, 16: 141–143.

Sharaf-Eldin, M., Elkholy, S., Fernandez, J., Junge, H., Cheetham, C., Guardiola, J. and Pamela. 2008. Bacillus subtilis FZB24® affects flower quantity and quality of Saffron (*Crocus sativus*). Planta Medica, 74:1316-1320.

Sharma,T., Kaul, S. and Dhar, M, K. 2015. Diversity of culturable bacterial endophytes of saffron in Kashmir, India. Springer Plus,4: 661.

Shuab, R., Malla, N. A., Ahmad, J., Lone, R. and Koul, K. K. 2016. Arbuscular mycorrhizal fungal symbiosis with saffron (*Crocus sativus* L.) plant. Journal on New Biological Reports, 5: 59 – 67.

Stapleton, J. J. and DeVay, J. E. 1995. Soil solarization: a natural mechanism of integrated pest management. In: Reuveni, R. (eds.). NovelApproaches to Integrated Pest Management. pp 309–322, Lewis Publishers, Boca Raton, Florida, USA, 1995.

Sud, A. K., Paul, Y.S. and Thakur, B. R. 1999. Corm rot of saffron and its management. Journal of Mycology and Plant Pathology,29: 380-382.

Thakur, R. N., Singh, C. and Kaul, B.L. 1992. First report of corm rot in Crocus sativus L. Indian Phytopathology, 45: 278-282.

van de Mortel, J. E., de Vos R. C. H., Dekkers, E. Pineda, A., Guillod, L., Bouwmeester, K, van Loon, J. J. A., Dicke, M. and Raaijmakers, J. M. 2012. Metabolic and transcriptomic changes induced in Arabidopsis by the rhizobacterium Pseudomonas fluorescens SS101. Plant Physiology, 160: 2173-2188.

Wani, Z. A., Ahmad, T., Nalli, Y., Ali, A., Singh, A. P., Vishwakarma, R. A., Ashraf, N. and Riyaz-Ul-Hassan, S 2018. Porostereum sp., associated with saffron (*Crocus sativus* L.), is a latent pathogen capable of producing phytotoxic chlorinated aromatic compounds. Current Microbiology, 75:880-887.

Wani,Z. A., Kumar, A., Sultan,P., Bindu,K., Riyaz-Ul-Hassan, S. and Ashraf, N. 2017. MortierellaalpinaCS10E4, an oleaginous fungal endophyte of Crocus sativus L. enhances apocarotenoid biosynthesis andstress tolerance in the host plant. Scientific Reports, 7: 8598.

Whipps, J.M. 2001. Microbial interactions and biocontrol in the rhizosphere. Journal of Experimental Botany, 52:487–511.

Winterhalter, P. and Straubinger, M. 2000. Saffron-renewed interest in an ancient spice. Food Reviews International, 16: 39–59.

Xu, C.X. and Ge, Q.X. 1990. A preliminary study on corm rot of *Crocus sativus* L. Acta Agriculturae, Universitatis Zhejiangensis, 16: 241-246.

Yamamoto, W., Omatsu, T. and Takami, K. 1954. Studies on the corm rots of Crocus sativus L. on saprophytic propagation of Sclerotinia gladioli and Fusarium oxysporumf. sp. gladioli on various plants and soils. Scientific Reports Hyogo University Agriculture. 1: 64–70.

3

Opportunities and Scope of Farm Mechanization in Himalayan Region

Manpreet Kour[1], B. C. Sharma[1], Jyoti Kachroo[2] and Prakarti Malakar[1]

[1]Division of Agronomy, FOA, SKUAST Jammu, Jammu and Kashmir, India
[2]Division of Agricultural Economics and ABM, FOA, SKUAST Jammu, India

Abstract

Jammu and Kashmir is a hilly state which is suitable for cultivation of a wide range of agri-horticultural crops with a great potential for development. Various studies indicate that in J & K, area, production and productivity of major food crops is increased, but the rate of increasing is slow. The present agricultural production is insufficient to fulfill the need of growing population due to the limited scope of increase in cultivable area and low level of mechanization. It was observed that mechanization in agriculture sector is not up to mark because the main problems faced by the farmers in adoption of improved farm equipment. Indian farmers were lack of awareness about various farm implements and machinery. The productivity of farms mainly depends on the availability and proper use of farm power by the farmers. Agricultural machines increase productivity of land and labour by meeting timeliness of farm operations and increase work out-put per unit time. Beside it multiple cropping and diversification is possible in agriculture through mechanization. Effective utilization of seeds, fertilizers and irrigation water it is necessary to adoption of mechanization technology. The level of farm mechanization in J&K is very poor due to its hilly area. Mechanical power, efficient implements, water management, land reclamation and post-harvest technology sectors have very much possibility of growth. Several studies suggest a direct correlation between farm mechanization and crop productivity. The farm mechanization is poor in India and Jammu & Kashmir due to small and irregular fields, undulating topography, lack of skilled manpower, poor facilities of repair, maintenance and manufacture of implements and high cost solar gadgets. Farm mechanization saves inputs like seeds 15-20%, fertilizers 20-30%, labour 5-20% and operational time by 20–30% and increases cropping intensity by 5–20% and crop productivity by 10–15%.Non-availability of improved equipment, small and scattered land holding, low investing capacity of the farmers and lack of awareness among farmers are the main bottleneck in mechanization. To strengthen the agricultural engineering wing in Jammu and Kashmir is required to attention of the state government and other funding agencies. A cell of agricultural engineering should be established to handle the farm mechanization problems. Despite various limitations, there is a great scope for increasing productivity of land and farmer's economy through land development, use of efficient farm power and implements.

Keyword: Mechanization, Farm Equipment, Crop productivity, Machinery

Introduction

Agriculture sector has the most important role in Indian economy. Earlier, Indian farmers largely depend upon the human and animal power. But with time passage, implements viz. tractor and tractor driven have been introduced. During the last 10 years, It has been observed that the concept of farm mechanization and its importance has been well accepted and accordingly new farm implements/ equipments have been developed and commercialized. This has resulted in increase of productivity and production to feed the ever increasing population besides reduction in drudgery associated with farm operations. Farm mechanization has played a positive role in increasing employment in rural areas through generation of opportunities for operators, mechanics, salesmen etc. (Anonymous, 2018)

Indian agriculture has capability of producing most of the food and horticultural crops of the world. It has good ranking in production of a number of crops viz. rice, wheat, sugarcane, fruits and vegetables. Major issues of Indian agriculture are the stagnancy in productivity and shortage of agricultural produce. Several studies suggest a direct correlation between farm mechanization and crop productivity. It saves inputs like seeds 15-20%, fertilizers 20-30%, labour requirement reduced 5-20% and operational time by 20–30% (Singh *et al.*, 2008; Ahmed *et al.*, 2020). Cropping intensity increases by 5–20% and crop productivity by 10–15%. Now, the farm mechanization adoption rate in present time is faster in Indian farmers, as compare to recent past. Tractors and power tillers are fulfilling the high demand of mobile power for timely farm operation and increased cropping intensity. Number of tractors in India has increased from 0.037 million in 1960–61 to 4.427 million units in 2010–11. It is estimated that number of tractors will 20.36 million in India by 2032–33. The availability of farm power from the tractors has accordingly grown from 0.007 kW/ha in 1960–61 to 1.03 kW/ha in 2013–14 at an overall growth rate of about 10% in the last 54 years. It is estimated that power availability from tractor will reach to 3.74 kW/ha by 2032–33. (Tiwari *et al*, 2019).

Over the years, the farm power has been shifted towards the mechanical and electrical sources. Animate power sources were contributed 93% intotal farm power in the year 1960–61, which has reduced to about 12.6% in 2010–11 (Fig 1), whereas in the same period mechanical and electrical sources of power have increased from 7% to 87.4%. It is estimated that in the year 2032–33 animate power will remain only 4.1% in total farm power (Tiwari *et al*, 2019).

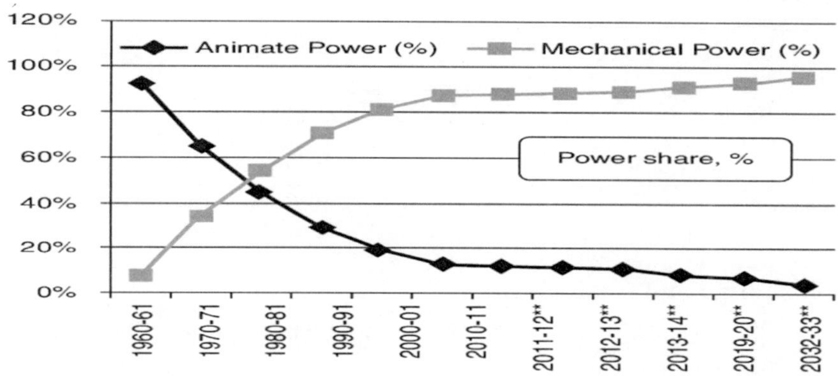

Fig 1. Trends of animate and mechanical power in Indian agriculture
Source: Singh and Sahni (2019) and Tiwari *et al*, 2019

Farm Mechanization

The process of developing agricultural machines and substituting this machine power for human and animal power in agricultural production practices is farm mechanization.

Mechanization of farms means the use of machines for conducting agricultural operations, replacing the traditional methods which involve human and animal labour. Farm mechanization implies the use of mechanical technology in the varied farming operations like sowing, harvesting, threshing, levelling, watering, spraying, weeding etc.

Need of farm mechanization

- To increase the production and productivity of food grains and to meet the global standards.
- For timely operations of agricultural activities.
- Efficient utilization of inputs, fertilizers, water and other natural resources.
- To meet the growing needs of growing population in almost the same amount of cultivable land.

- Availability of adequate farm power is very crucial for timely farm operations for increasing production and productivity and handling the crop produce to reduce losses.
- Efficient machinery helps in increasing productivity by about 30% besides, enabling the farmers to raise a second crop making the agriculture attractive.
- Farm mechanization saves inputs

Seeds by	15–20 %
Fertilizers by	15–20%
Labour requirement by	20–30%,
Operational time by	20–30%,
Increases cropping intensity by	5–20%
Crop productivity by	10–15%
Increase in gross income of the farmers	29-49

A View of Jammu and Kashmir

Jammu & Kashmir is located between 32°17' to 36°58' North latitude, and between 37°26' to 80°30' East longitude. It is covering an area of 2,22,236 sq. km and it is The state is generally known as Kashmir. Agriculture occupies an important place in the economy of the state. The share of Agriculture and allied sectors in the Gross State domestic Product as per Advanced estimates stands at 20.59% for the year 2010-11. On the other hand nearly 70% of the population in the state derives its lively-hood directly or indirectly from the agriculture sector (Anonymous, 2017-2018). Agriculture is and would continue to be the major source of income and employment for the people of State, due to limited resource endowments in other sector. The agriculture and its allied sectors provide about 36% of the net domestic product and engage about 80% of the population. Draught animals are the main source of farm power. Indigenous, Tawi, Hill and Soil stirring plough areanimal drawn implementsused for land preparation activities. Manual tillage implements such as spades, shovels, sickle, hand forks areprobably used by small and marginal farmerswhile farmers lived in plain topography use tractors and implements for tillage and transportation purpose (Dixit *et al.*, 2006, Sharma, 2010). As adoption of agricultural machinery increases capital input, it is expected that the level of output should increase. Such changes are likely to have an impact on employment, income and income distribution and social welfare (Shambhu and Jha, 2012).

Farm mechanization level in State is low. In the state only 41.7% of the net sown area is irrigated by different sources and remaining area (58.3%) is rainfed. The major source of irrigation is canals (90.01%). There are 13.35 lakh land holdings in the State with a population of 101.44 lakh. Nearly 82% of the total geographical area is not available for cultivation whereas rest 18% area is being used for agriculture and other allied activities of which only 7.36% is the net sown area.

Net sown area has increased from 0.614 million ha to 0.736 million ha during the last more than 60 years (Raina *et al,* 2018). It is estimated that 0.35 million t food grain is required to ensure food and nutritional security of State population. So it is necessary to increasing production and productivity of food grain and for this purpose is required to adaptation of mechanization in agriculture and allied sector.

Seed Bed Preparation (Tillage) Equipment

Mould-board plough - It is used for primarytillageoperationsorforopeningup any cultivable land.

Rotavator -it is used for seedbed preparation, weed control, mixing of soil with crop residue and fertilizer and pudding of the soil. Using it cut the time (30-35%), water (30%) and cost of operation (20-25%) as compared to tillage by cultivator and harrow. Good seedbed and pulverization of soil isachieved by a single pass.

Tractor drawn Disc harrow–It issuitableforhard ground, full of stalks and grasses. It cuts the lumps of clods, soil and roots. Discs are mounted on one, two or more axles, as it is pulled ahead, discs rotate on the ground.

Laser land leveller -It is leveling the field within certain degree of desired slope using a guided laser beam throughout the field. It contains a laser guided scrapper, transmitter and receiver. It is a precision equipment to level the field (Sundaram *et al,* 2012).

The laser land levelling significantly affected the yield and yield attributes of rice and wheat crop (Table 1). Laser leveled field show maximum plant hight, number of panicle m^{-2}, grain yield and water use efficiency of rice and wheat against unlevelled field.

Table 1: Yield attributes and yield as affected by levelling techiniques in Rice-Wheat system

Treatments	Plants height (cm.)		Effective tiller m^{-2}		No. of grain spike^{-1}		1000 grain wt.(g)		Grain yield (t. ha^{-1})		WUE (kg m^3)	
	Rice	Wheat	Rice	Wheat	Rice	Wheat	Rice	Wheat	Rice	Wheat	Rice	Wheat
Laser land leveling	128.7	96.4	318	325	132	42.4	24.3	41.7	4.70	4.91	0.42	1.48
Traditional leveling	123.8	88.2	302	310	121	39.1	23.5	38.6	4.11	4.43	0.32	1.17
Control (Unlevelled)	110.5	79.4	287	289	115	37.4	21.2	37.6	3.68	4.14	0.26	0.98
CD(P=0.05)	14.6	11.2	14.2	16.0	6.5	3.8	1.4	3.1	0.26	0.22	-	-

Source: Tomar*et al*, 2020

Table 2. Grain yield (kg ha^{-1}) as affected by different tillage practices and herbicide application during wheat growingseason.

Tillage	Year		2-years Mean
	2017-18	2018-19	
MB Plough (T1)	4356.4	4406.4	4381.4
Disc Harrow (T2)	4187.6	4237.6	4212.6
Rotavator (T3)	4340.9	4390.9	4365.9
Cultivator (T4)	4127.5	4177.5	4152.5
LSD for Tillage	86.42	86.42	56.75
Herbicide			
Pre-Emergence (H1)	4409.1	4459.1	4434.1
Post-Emergence (H2)	4311.3	4361.3	4336.3
Control (H_o)	4039.0	4089.0	4064.0
LSD for Herbicide	45.026707	45.026707	31.017
Year			
2017-18			4253.1
2018-19			4303.1
Significance			*
Interaction	Significance	Interaction	Significance
Y x T	NS	Y x H	NS
T x H	*	Y x T x H	NS

Source: Uddin and Amin, 2019

Uddin and Amin, 2019 find out that different tillage implements, herbicides application and interaction between TxH had significant effect on grain yield presented in Table 2. Maximum grain yield (4381.4 kg ha^{-1}) was recorded in the plot prepared by mould board plough followed by rotavator. On the other hand, significance interaction effect showed by TxH. According it improved tillage practices with herbicide application give higher yield.

Sowing and Fertilizer Application

Zero till seed cum fertilizer drill - It helps to seed a crop directly into the cultivated field just after the harvest of the previous crop with the least disturbance of the soil. It eliminates or reduces time and energy intensive conventional tillage operations. The furrow openers are of inverted 'T' type. Use of zero-till drill for direct sowing of wheat after rice saves time and cost as compared to the conventional practice of seedbed preparation and sowing.

Seed cum fertilizer drill - Seed cumfertilizer drill isused in the same way as zero till seed cum fertilizer drill, but in the cultivated land i.e.,after the seed bed preparation.

Dibbler - It is single row manually operated equipment for dibbling bold or medium seeds in row or gap filling into well prepared soil. It is suitable for drilling wheat, field pea, and maize in small plot.

Happy Seeder -It combines two units, one for straw management and the other one is for sowing wheat after paddy. Green gram can also be sown in summer after wheat harvest in standing stubble which helps in moisture conservation and soil temperature regulation (Sundaram *et al,* 2012).

Table 3. Performance of seed cum fertilizer drill and normal seed drill on Soyabean

Economic parameters	Two year pool data for Soybean			
	Ridge and furrow seed cum fertilizer drill	Normal Seed drill	% increase over seed drill	CD at 5%
Plant height at harvesting (cm)	62.6	54.2	15.50	S
Number of Branches per plant at 60 DAS	5.57	5.22	6.70	S
Number of root nodules per plant at 60 DAS	29.4	22.6	30.09	S
Number of pods per plant at harvesting	42.8	31.9	34.17	S
Seed Index (g)	12.62	12.43	1.53	NS
Grain yield (kg/ha)	1309	1091	19.98	S
Straw yield (kg/ha)	1606	1362	17.91	S
Grain straw ratio	0.824	0.799	3.13	NS
Harvest index (%)	44.9	45.1	-0.44	NS
Net monetary returns (Rs/ha)	25144	18025	39.50	S
Benefit: cost ratio	2.43	2.11	15.17	S

Source: Dhakad *et al,* 2019

Ridge and furrow seed cum fertilizer drill was higher the grain yield (1309 kg ha^{-1}), straw yield (1606 kg ha^{-1}) and net monetary return (Rs 25144) compare to normal seeddrill sowing (Table 3). It is showed that using ridge and furrow seed cum fertilizer drill in showing give maximum benefit than normal seed drill sowing.

Table 4. Growth and yield of the wheat crop as influenced by different planting methods.

Sowing Technology	Plant height (cm)	Effective Tillers/m^2	Ear length (cm)	No. of grains/ ear	1000 grain weight	Grain yield (q/ ha)	Straw yield (q/ ha)
Conventional method	90.35	439.50	11.01	60.85	42.5	52.74	73.80
Incorporation (MB plough + disc harrow + rotavator)	91.60	441.50	11.10	61.05	43.2	53.20	74.84
Happy Seeder	90.12	440.90	10.95	60.77	42.95	52.60	72.67
Zero tillage	88.75	425.60	10.42	59.42	42.66	51.10	69.85
Sig. (p = 0.05)	0.141	0.000	0.844	0.759	0.869	0.396	0.25

Source: Gautam *et al,* 2020

Gautam *et al,*2020 reported thatthe effective tillers per square meter were significantly higher among the planting method of happy seeder (440.90), which was statistically at par with the conventional method (439.50). They reported that crop residue is big problem in the rice-wheat cropping system for its management happy seeder will be provided a better option. Rice straw contains various nutrients in significant quantities; their continuous application will have a positive outcome on fertilizer management in the rice-wheat system. Table 4 showed that grain and straw yield obtained by happy seeder was significantlyatparwithconventionalandincorporationtillage,in the long duration, use happy seeder increase yield due to the above-mentioned reason.

Table 5. Yield and Economics of the wheat crop as influenced by different planting methods.

Sowing Technology	Yield (q/ha)	Input Cost (Rs)	Return of grain (Rs)	Gross return of straw (Rs)	Gross return (Rs)	Net Return (Rs)	B:C ratio
Conventional method	52.74	34517	91504	23616	115120	80603	2.34
Incorporation (MB plough + disc harrow + rotavator)	53.20	35023	92302	23949	116251	81228	2.32
Happy seeder	52.60	32517	91261	23254	114515	81998	2.52
Zero tillage	51.10	32017	86924	21072	107996	75979	2.37

Source: Gautam *et al,* 2020

Between the different types of sowing technologies the incorporation method (Rs.116250.80) give highest gross return followed by the conventional method (Rs. 115119.90), Happy seeder (Rs.114515.40),No-till drill/zero tillage, respectively (Table 5).The benefit-cost ratio was also High with happy seeder (2.52:1) as compared to incorporation (2.32:1) and other methods (Table 5). High B: C ratio with happy seeder was due to its lowcost of cultivation as compared to incorporation and conventional method of sowing.

Table 6. Performance of different types of weeder in wetland Rice

Weeding method	Weed population m^{-2}		Weeding efficiency (%)	Field efficiency (%)	Energy consumption (MJ/ha)	Cost of operation (Rs/ha)
	No. of weed before weeding	No. of weed after weeding				
1. Hand weeding	222	10	95.28	-	927.16	4250
2. Ambika paddy weeder	182	40	78.03	69.45	270.40	1050
3. Conoweeder	168	47	70.90	68.42	302.11	1160
4. Power operated rotary weeder	245	22	91.00	61.00	1521.80	808.42

Source: Netam and Mahilang, 2018

According to Table 6 that Ambika paddy weeder showed maximum field efficiency (69.45%) due to higher cut of width and less clogging in soil compared to other treatment. Higher energy consumption in Power operated rotary weeder due to involvement of man and machine. Power operated rotary weeder show minimum cost of operation (Rs/ha 808.42) and highest cost of opration observed under hand weeding (Rs/ha 4250). It is show that using Power operated rotary weeder's cost of opration can be minimized due to less time engagement of human labourers as compared to other weeding method.

Table 7. Effect of irrigation methods on growth, yield attributes and water productivity of wheat crop

S.No.	Parameters	Irrigation method	
		Borders irrigation	Pressurized Sprinkler
1	Plant height (cm)	84.40	88.56
2	No. of tillers /m^2	382.5	396.8
3	Ear length (cm)	8.05	9.10
4	1000 grain wt. (gm)	40.12	44.28
5	Biological yield (q/ha)	90.28	104.11
6	Grain yield (q/ha)	39.50	45.91
7	Straw yield (q/ha)	50.78	58.20
8	Total irrigation water applied (cm)	45.00	40.00
9	Water productivity (kg/ha/cm)	87.77	114.77

Source: Sharma *et al,* 2018

Table 7 presented that highest grain yield and straw yield was observed under pressurized sprinkler method. It also revealed that 30.76% more water productivity in sprinkler irrigated wheat compared to surface border irrigated wheat. It show that using pressurized sprinkler method in wheat can save 12.5% water and 16.22 % yield can increased compared to surface boarder irrigation method.

Harvesting and Thrashing Equipment

Multi crop thresher - Suitable to thresh crops like wheat,maize, sorghum, gram, soybean, pigeon pea and sunflower, Operated by 20 hp three phase electric motor or 35 hp tractor pto (power takeoff). It consists of hopper with automatic feeding, spike tooth cylinder, three aspiration blower, concave, shaker unit, bagging unit, hitch system and pneumatic transport wheel.

Tractor operated straw reaper or straw combine - Straw combine is used to recover wheat straw after combine operation and is operated by a tractor. Straw collectedby straw combine is cut into pieces and collected in the trolley. Also some grains are collected alongwith straw.Thecapacity of machine on an averageis 0.4 hah^{-1}while operating at a speed of 2.5 kmh^{-1}andstraw recovery is about 55-60%.

Two persons are required for its operation. The quality of straw is comparable with mechanical thresher. There is also additional grain recovery of 50-100 kg ha^{-1}. The cost of grain recovered can be compensated tothe amount paid for hiring themachine.

Tubular maize sheller - It is a hand operated tool to shell maize from dehusked cobs. The unit consists of galvanized mild steel pipe with four tapered fins riveted to its inner periphery, the Sheller is held in left hand, a cob held in right hand is inserted into it with forward and backward twist, to achieve the shelling. Octagonal designs are also available (Sundaram *et al,* 2012).

Table 8. Effect of manual harvesting and mechanical threshing and combine harvester in threshing efficiency (%) of rice varieties

Treatments (T) Varieties (V)	Threshing efficiency (%)			
	Manual harvesting and threshing	Manual harvesting and mechanical threshing	Combine harvesting	Mean
CO51	99.323 (85.30)	99.709 (86.94)	99.985 (89.26)	99.67 (87.17)
CR1009 Sub 1	99.638 (86.59)	99.750 (87.21)	99.995 (89.40)	99.79 (87.73)
Improved white ponni	99.663 (86.72)	99.827 (87.67)	99.989 (89.32)	99.83 (87.90)
Mean	99.54 (86.20)	99.76 (87.27)	99.99 (89.33)	99.76 (87.60)
	V	T		VxT
SEd	0.268	0.268		0.464
CD (P=0.05)	0.567	0.567		NS

Source: Govindaraj *et al,* 2017

Result showed in table 8 between different varieties, the highest threshing efficiency shows in improved white ponni (99.83%) and CR 1009 Sub 1 (99.79 %) while among the different harvesting and threshing methods, combine harvesting method (99.99%) showed maximum threshing efficiency. It revealed that improved threshing methods with improved variety give maximum threshing efficiency in less time.The highest broken seed was observed in seeds harvested and threshed by combine harvester (3.53 per cent) followed by manual harvesting and mechanical threshing (0.69 per cent) (Table 9).

Table 9. Effect of manual harvesting and mechanical threshing and combine harvesting on broken seed (%) of rice varieties

Treatments (T) Varieties (V)	Broken seed (%)			
	Manual harvesting and threshing	Manual harvesting and mechanical threshing	Combine harvesting	Mean
CO51(V_1)	0.016 (0.723)	0.841 (5.252)	4.428 (12.131)	1.762 (6.035)
CR1009 Sub 1	0.012 (0.630)	0.536 (4.180)	2.668 (9.358)	1.072 (4.723)
Improved white ponni	0.019 (0.789)	0.692 (4.678)	3.483 (10.569)	1.398 (5.345)
Mean	0.016 (0.714)	0.690 (4.703)	3.526 (10.685)	1.411 (5.368)
	V	T		VxT
SEd	0.498	0.498		0.863
CD (P=0.05)	NS	1.047		NS

Source: Govindaraj *et al,* 2017

Government schemes

Multiple initiatives have been taken by Government of India to improve farm mechanisation in the country. Various farm mechanisation programmesand schemeshave been launched by Government of India across the country, covering various aspects of incentives. Some of these important schemes are:

The Sub Mission on Agricultural Mechanization (SMAM), a centrally sponsored scheme launched in the year 2014-15 by the Ministry of Agriculture and Farmers Welfare for promotion of farm mechanization (Anonymous, 2018-19). The main components of the scheme are;

- Training, testing and demonstrations of farm machinery and equipment
- Establishment of custom hiring centre
- Subsidy through state departments of agriculture for purchase and distribution of farm implements to the farmers
- Promotion of post-harvest technology for encouraging primary processing.

Scheme	Support for farm mechanisation	Funding/financing pattern
NFSM	Assistance (up to 50% the cost of machinery) to be provided for machinery such as pump sets, tractor mounted sprayers, seed drills, zero till seed drills to varying degrees.	60% Centre and 40% state
RKVY	• Farm mechanisation comes under production growth stream of RKVY – with 35% of the outlay. • Assistance for large equipment (tractors, combine harvesters, sugarcane harvesters, cotton pickers, etc.) is available for establishing custom hiring centres (CHCs)	60% Centre and 40% state
MIDH	• One of the key interventions under the scheme is 'horticulture mechanisation'. • Assistance is provided for procurement of power-operated machines and tools, besides for import of new machines.	60% Centre and 40% state

Source: FICCI and PwC knowledge report, 2019.

It was found that the average Farm Power Availability (FPA) at the end of 2013-14 was 1.84 kW/ha, whereas after three years of SMAM implementation, it has risen to 2.025 kW/ha. Thus, an increase of 17.3 % has been recorded. The future requirement has been worked out at 2.50 kw/ha during the year 2022 and 4.0 kw/ha during year 2030 (Anonymous, 2018-19).

Problems relating to farmers

1. Small land holdings size and scattered land a big problem.
2. Financial inability of the small farmers.
3. Lack of knowledge about farm machineries and equipment.
4. Lack of repair and replacement facilities especially in the remote rural areas.
5. Limited availability of sale outlets.
6. Lack of accessibility for procurement, repair and maintenance.

Suggestion

The farm machinery suitable for different types of soil and operations for important crops should be developed and manufactured.

Technical know-how should be provided to the farmers with respect to appropriateness of farm machinery for the situation and for its proper use.

A standardization and quality-marking centres of farm equipment should be established in potential areas.

Conclusion

1. In Indian agriculture, power share percent under mechanization has increased enormously. Farm mechanization saves inputs (15-20%), labour& operation time (20-30%), cost of production (20-25%) besides enhancing WUE, crop intensity (15-20%) and productivity(10-15%).
2. Mechanical equipment which have higher output capacity and cut down the number of operations to be performed, have helped in increasing area under cultivation and consequently lead to increase in cropping intensity.
3. For, efficient utilization of inputs, fertilizers, water and other natural resources and to meet the needs of growing population, availability of adequate farm power is very crucial.
4. Rice-wheat cropping system has huge potential of food security, therefore, this system needs to mechanized for higher production.
5. Laser land levelling has shown to improve water management, crop stand and productivity. With laser levelling, significant savings in irrigation water, improved efficiency of inputs, higher yields, improved soil health, increase in cultivated area, and better operational efficiency can be achieved.
6. Happy seeder technology provided the facility of wheat sowing against the burning of paddy straw. The technology is eco- friendly, as well as time saving.

References

Ahmed, S., Bagal, Y.S., Mahajan, R. and Sharma, L.K. 2020. Impact of Farm Mechanization on Crop Productivity in Sub- Tropical Areas of Jammu and Kashmir. Int.J.Curr.Microbiol. App.Sci., 9(05): 1168-1173.

Anonymous. 2017-18. Digest of Statics of Jammu and Kashmir. Dept. of Statics, Jammu and Kashmir, India.

Anonymous. 2018-19. Operational Guidelines, Sub- Mission on Agricultural Mechanization.

Anonymous. 2018. Monitoring, Concurrent Evaluation and Impact Assessment of Sub-Mission on Agricultural Mechanization. Ministry of Agriculture & Farmers Welfare (Department of Agriculture, Cooperation & Farmers Welfare) Mechanization & Technology Division.

Dhakad, S.S., Ambawatia, G.R., Singh, M. and Verma, G. (2019). Effect of Ridge and Furrow Seed cum Fertilizer Drill on Growth Characters and Yield of Soybean (Glycine max) in Shajapur District of Madhya Pradesh, India. Int.J.Curr.Microbiol.App.Sci. 8(04): 2298-2304.

Dixit, J., Jeena, A.S., Shahi, N.C. and Wahid, T. (2006). Potential of Farm Mechanization in Jammu and Kashmir State of India - a Review. Agricultural Mechanization in Asia, Africa, and Latin America37(4): 9-14.

FICCI and PwC knowledge report, 2019. Farm mechanisation: Ensuring a sustainable rise in farm productivity and income.

Gautam, A., Singh, V. and Aulakh, G.S. (2020). Effect of various sowning technology of wheat cultivation under rice-wheat cropping system in the western plain zone of Punjab. Multilogic in Science,32(9): 459-461.

Govindaraj, M., Masilamani, P., Asokan, D., Rajkumar, P. and Selvaraju, P. (2017). Effect of Different Harvesting and Threshing Methods on harvest Losses and Seed Quality of Rice Varieties. Int.J.Curr.Microbiol.App.Sci. 6(9): 1510-1520.

Netam, A. and Mahilang, K.K.S. (2018). Comparative Performance of Different Types of Weeder for Wetland Rice in Chhattisgarh Region, Int. J. Pure App. Biosci. 6(1): 243-245.

Raina, S., Dadhich, H., Kumar, A., Singh, B. and Kumar, J. (2018). Status, Scope and Constraints of Farm Mechanization in Jammu and Kashmir State of India. Int.J.Curr.Microbiol.App. Sci7(3): 1279-1286.

Shambhu, V.B. and Jha, S.K. (2012). Problems and Prospects of Agricultural Mechanization in Bihar, India. Agricultural Mechanization in Asia, Africa, and Latin America 43(3): 55-59.

Sharma, S.K., Mishra, P.K., Panse, R. and Jamliya, G. (2018). Effect of Irrigation Methods on Yields Attributes and Water Productivity of Wheat in Vertisol of Betwa River Basin Commands of Vidisha District of M.P., India Int.J.Curr.Microbiol.App.Sci. 7(08): 2670-2673.

Sharma S. 2010. Status of farm power availability in the selected villages of Udhampur District (J&K State). Journal of Dryland Agriculture Research & Development, 25(1): 70-73.

Singh., et al., 2008, original not seen, cited at page no. 63 & 64 in state of Indian agriculture 2011-12.

Singh R S and Sahni R K. *2019.* Transformation of Indian agriculture through mechanization. Economic Affairs 64(2): 297–303.

Sundaram, P.K., Singh, S.S., Sharma, S.C. and Rahman, A. (2012). Prospect of Farm Mechanization. Status of Agricultural Development in Eastern India, 279-292.

Tiwari, P.S., Singh, K.K., Sahni, R.K. and Kumar, V. (2019). Farm mechanization – trends and policy for its promotion in India. Indian Journal of Agricultural Sciences89(10): 1555–62.

Tomar, S.S., Singh, Y.P., Naresh, R.K., Dhaliwal, S.S., Gurjar, R.S., Yadav, R., Sharma, D. and Tomar, S. 2020. Impacts of laser land levelling technology on yield, water productivity, soil health and profitability under arable cropping in alluvial soil of north MadhyaPradesh. Journal of Pharmacognosy and Phytochemistry, 9(4): 1889-1898.

Uddin, S. and Amin, M. (2019). Comparison of mechanical and chemical weed control in wheat- maize cropping system. Pak. J. Weed Sci. Res., 25(4): 303-313.

4

Livestock and Its Impact on the Environment

Vijay Kumar Sharma*, Neerja Sharma, Abhay Kumar Sinha and Saurav Gupta

KVK, Samba, SKUAST-Jammu, Jammu and Kashmir, India

Introduction

India is often criticized for the negative externalities to the ecology due to the largest livestock population of the country. Thus inspite of the fact that the integrated farming system followed in the country contributes towards not only saving the resources and improving the environment. This study quantifies the positive environmental externalities associated with the livestock production in the context of integrated farming system of rainfed areas. It included land saving due to recycling of agricultural by-products as animal feed and also due to use of dung- cake as domestic fuel, saving of chemical fertilizers due to use of dung as manure and prevention of carbon dioxide emission due to use of animal energy in agriculture.

India being one of the largest livestock keeper of the world, its livestock sector comes under the stringent scrutiny of the global climate monitoring agencies such as IPCC. This inspite their significant contributions towards enhancing food and nutritional security are often criticized for the negative externalities they cause to environment through emission of greenhouse gases, overgrazing/ deforestation and water pollution. According to the estimates of the Indian Network for Climate Change Assessment, the agricultural sector emitted 334 Mt of CO_2 in 2007, to which livestock contributed about 63 per cent. Further, livestock contribute 18 percent of global greenhouse gas (GHG) emissions (Steinfeld *et al.,* 2006). Apart from enteric emission decomposition of livestock manure under anaerobic conditions is alsoa source of methane. Another green house gas is nitrous oxide (N_2O) released during the nitrification-denitrification of nitrogen contained in livestock waste. Cattle and feedlots are responsible for 26% of N_2O emissions from anthropogenic sources (IPCC, 2001). Being green house gases their large scale emission is detrimental to the environmental safety. So, different strategies are emerging to either subside their emission from faeces and animal wastes or to use them effectively for energy saving purposes. A number of studies in the past have quantified methane emission from livestock production. These studies have shown wide variations, ranging from 8.5 Mt to 10.5 Mt, depending on the methods and data used and also the year for which the gas emission was estimated (Lerner *et al.*, 1988; Ahuja, 1990; Bhattacharya and Mitra, 1998; Mishra and Dikshit, 2004; Singhal *et al.*, 2005; Swamy and Bhattacharya, 2006; Singh *et al.*, 2012). Negative externalities of livestock to environment are well documented and quantified (Steinfeld *et al.*, 2006) but their positive contributions have remained less documented and non-quantified. The prominent positive environmental contributions include prevention of carbon dioxide emission due to use of animal energy in place of fossil fuel, saving of natural resources mainly land as a result of recycling of agricultural by-products and residues as animal feed, and dung in place of firewood as domestic fuel and as a substitute for chemical fertilizers.

Estimation of Feed Consumption Rates

Household level feed consumption rates serve as a base to estimate feed consumption rates at the national level. From the survey, we gathered information on (i) number of sample households having livestock, say cattle in milk and (ii) number of cattle in-milk observed,and (iii) amount of feed fed per day to these cattle in-milk. We then scaled-up information.

Utilisation of land saving because of recycling of crop residues as animal feed

By using the feed consumption rates, we can quantify the positive contributions of livestock production systems to the environment following the "Environmental Model of Livestock Production System" by Mishra and Dikshit (2004). The estimated positive impact of the system included land saving due to crop by-product used as animal feed, and due to use of dung as a domestic fuel; saving in chemical fertilizers due to dung-use as a manure; saving in fossil fuel (diesel) due to use of animal energy in agricultural operations. The model has been described as follows:

The gross energy intake per animal per day from by-product feed was estimated by summing up the energy values of by-product feed on dry matter basis. Similarly, the energy value of green fodder fed to the animals was also calculated on dry matter basis. Then, the annual quantity of green fodder required, in terms of energy to replace grossenergy from by-product feed, was estimated as per equation (1) :

$$G_t = \left[\frac{1}{e} \times gei \times \frac{1}{d} \right] \times 365$$

where, Gf is the quantity of green fodder required to replace the by-product feed, gei is the gross energy intake from by-product feed (dry fodder and concentrates), e stands for the energy (million calories) per unit of Gf; and d is the dry matter fraction of green fodder. We then estimated the land area required to produce Gf. Let y be the yield of green fodder per hectare of land, then the area L required to produce Gf may be given by equation (2):

$$L = \frac{G_t}{y} \qquad (2)$$

Land Saving due to Use of Dung as Substitute for Fire-wood in domestic fuel

In rural households, fire-wood is used as a domestic fuel; for cultivation or perennial fire-wood trees cover land and deprive its usefor farming. The use of dung cake as fire-wood would result in the saving of this land, which can be used for crop-cultivation. We, therefore, estimated the land saved due to use of dung cake as a domestic fuel. The dung-cake output on dry matter basis was workedout using dung evacuation rate and its dry matter fraction. Supposing as the rate of substitution of dry fuel-wood for dung cake (fuel wood: dry dung cake) in terms of thermal energy, the total quantity of dry fuel-wood required to replace the supply (output) of dung cake was calculated by equation (3):

$$F_w = \delta \times dc \qquad (3)$$

where, Fw is the quantity of fuel-wood required to replace dung cake, and dc is the quantity of dung cake output. Now, let us suppose that fuel-wood is produced and used within the year, and its yield per hectare on dry matter basis is y, then the total land area that would be required for producing Fw can be calculated by equation (4):

$$L = \frac{1}{y} \times F_w \qquad \ldots(4)$$

The effect of gestation lag in the production of fuel wood can be described as follows: The model assumes that fuel-wood is produced and used within the same year. This is apparently an unrealistic assumption. In reality, more than one year is required to cover the whole process of fuel-wood tree plantation, growth and logging of trees, drying and use of cut-out wood as fuel. Suppose it takes 3 yearsto complete the process before dry wood is made available for use at the end of the 3rd or the beginning of 4th year. This means that whereas the fuel wood made available can replace equivalent amount of dung cake only in the fourth year, the necessary land area required for growing and harvesting of trees for making the fuel-wood available will have to be kept locked up during the preceding 3 years. This implies that around 3-times as much land will be required or saved if one year's dung cake output was used in place of fuel wood as energysource.

Saving of Chemical Fertilizers due to Use of Dung as Manure

The extent to which dung manure substitutes the chemical fertilizers, is a saving of chemical fertilizers. Fresh cattle dung on an average contains 0.30-0.40 per cent nitrogen (N), 0.10-0.20 per cent phosphorus (P_2O_5), and 0.10-0.30 per cent potassium oxide (K_2O) (Anonymous, 1997). According to the recent estimates of Ghosh *et al.* (2004), dung manure contains 0.71 per cent N, 0.18 per cent P_2O5 and 0.71 per cent K_2O. In this study, we have used the fraction of soil nutrients (N,P and K) in dung-manure as estimated by Ghosh *et al.* (2004), and the total quantity of N, P and K was worked out for the proportion of bovine dung used as manure.

Saving of Fossil Fuel due to Animal Energy-use in Agriculture

If the working animals were to be replaced by agricultural machinery, it would require additional fuel and lead to emission of CO_2 from burning of the fossil fuel. It is this emission that is preventedby the livestock. It has been assumed in the study that tractors are theonly machine used and are of similar power; and also the working animals do not differ in their work capacity. In order to determine thesaving of fossil fuel and prevention of greenhouse gasses due to animal energy use, we need to know (i) the rate of substitution between tractors and working animals, (ii) the amount of diesel required for running a tractor to perform agricultural operations, and (iii) the amount of CO_2 that will be emitted from the burning of a unit of diesel. On the basis of economic substitution rate, 10 working males are required to replace a tractor. The number of tractors required for replacing country's bovine working stock, diesel use, and the associated emission of greenhouse gases have been estimated. From time to time not only international organisations but also developed countries target India for its large livestock population. Thus a number of the studies were undertaken to quantify the methane emission from Indian livestock. These studies showed large variations ranging from 8.5 Mt to 10.5 Mt, depending upon the methods used as well as the period of the study (Lerner *et al.,* 1988; Ahuja, 1990; Bhattacharya and Mitra, 1998; Mishra and Dikshit, 2004; Singhal *et al.,* 2005; Swamy and Bhattacharya, 2006; Singh *et al.,* 2012). Inspite of the various negative externalities attributed, integrated farming system of the rainfed system contributes towards lowering the environmental impact.

Conclusion

Extension interventions have importance to reduce negative externalities of livestock besides popularising technologies for increasing productivity and further popularising animal diets that would result reduction in methane emissions.

References

Ahuja, D. (1990). United State Environmental Agency (USEPA), Inventory of U.S. Greenhouse Gas Emissions and Sink, 1990-98.

Anonymous (1997). Hand Book of Agriculture, Indian Council of Agriculture Research (ICAR), New Delhi.

Bhattacharya, S. and Mitra, A.P. (1998). Greenhouse Gas Emission in India for the Base Year 1990. Scientific Report, No.11, Centre for Global Change, National Physical Laboratory, New Delhi.

Ghosh, P.K. (2004) Comparative effectiveness of cattle manure, poultry manure, phosphor compost and fertilizer-NPK on three cropping systems in Vertisols of semi-arid tropics, Crop Yields and System Performance, 95: 77-83.

IPCC (Inter-Governmental Panel on Climate Change) (1995) Revised IPCC Guidelines for National Greenhouse Gas Inventories, OECD, Paris.

Lerner, J., Mathews, E. and Fung, I. (1988) Methane emission from animals, a global high resolution data, Global Biogeochemical Cycles, 2: 139-156.

Mishra, S.N. and Dikshit, A.K. (2004) Environment and Livestock in India : With a Comparative Study of the Indian and US Dairy Systems, Manohar Publishers and Distributors, New Delhi.

Singh, S., Kushwaha, B.P., Nag, S.K., Bhattacharya, S., Gupta, P.K., Mishra, A.K. and Singh, A. (2012) Assessment of enteric methane emission of Indian livestock in different agro-ecological regions, Current Science, 102(7): 1017-1027.

Singhal, K.K., Mohini, M., Jha, A. K. and Gupta, P.K. (2005) Methane emission estimates from enteric fermentation in Indian livestock: Dry matter intake approach, Current Science, 88: 119-127.

Steinfeld, H., Gerber, P., Wassennar, T., Castel, V., Rosales, M. and de Haan, C. (2006) Livestock's Long Shadow: Environmental Issues and Options, Food and Agriculture Organization of the United Nations, Rome.

Swamy, M. and Bhattacharya, S. (2006) Budgeting anthropogenic greenhouse gas emission from Indian livestock using country-specific emission coefficients, Current Science, 91: 1340-1353.

5

Importance of Biofertilizers in Agriculture

Saurav Gupta,Vinod Gupta, Neerja Sharma, A.K. Sinha and Vijay Kumar Sharma

Krishi Vigyan Kendra, Samba, SKUAST-Jammu, Jammu and Kashmir, India

Biofertilizers are products that contain microorganisms essential for soil fertility and plant growth when added to the soil. Biofertilizers consist of microbes and living bacteria that promote the fertility of the soil and the growth of plants. Biofertilizers are known to play a number of vital roles in soil fertility, crop productivityand production in agriculture as they are eco-friendly and cannot at any cost replacechemical fertilizers that are indispensable for getting maximum crop yields. Increasing useof chemical fertilizers in agriculture make country self-dependent in food production but itdeteriorate environment and cause harmful impacts on living beings. Due to insufficientuptake of these fertilizers by plant results, fertilizers reach into water bodies through rainwater, cause eutrophication in water bodies and affect living beings including growthinhabiting microorganism. The excess uses of chemical fertilizers in agriculture are costlyand also have various adverse effects on soils i.e. depletes water holding capacity, soil fertility and disparity in soil nutrients.

A number of micro-organisms (bacteria fungi and algae) are considered as beneficial foragriculture and used as biofertilizers. Biofertilizers are supposed to be a safe alternative tochemical fertilizers to minimize the ecological disturbance. Biofertilizers are cost effective,eco-friendly and when they are required in bulk can be generated at the farm itself. Theyincrease crop yield upto 10-40% and fix nitrogen upto 40-50 kg. The other plus point isthat after using 3-4 years continuously there is no need of application of biofertilizersbecause parental inoculums are sufficient for growth and multiplication. They improve soil1exture, pH.and other properties of soil. They produce plant growth promoting substances1AA, amino acids, vitamins etc. Biofertilizers contained 3.5% -4% nitrogen, 2% - 2.5%phosphorus and 1.5% potassium.

Important functions of biofertilizers in agriculture

- They supplement chemical fertilizers for meeting the integrated nutrient demand of the Crops.
- They can add 20-200 kg N/ha year (eg. Rhizobium sp. 50-100 kg N/ha year; Azospirillum, Azotobacter: 20-40 kg N/ha/yr; Azolla: 40-80 kg N/ha; BGA 20-30 kgN/ha/'yr) under optimum soil conditions and thereby increases 15-25 per cent of total cropyield.
- They can at best minimize the use of chemical fertilizers not exceeding 40-50 kg N/ha under ideal agronomic and pest-free conditions.
- Application of biofertilizers results in increased mineral and water uptake, root development, vegetative growth and nitrogen fixation.

Types of Biofertilizers

Rhizobium: Rhizobium is a soil habitat bacterium, which can colonize the legume roots and fixes the atmospheric nitrogen symbiotically. The morphology andphysiology of Rhizobium will vary from free-living condition to the bacteriod of nodules. They are the most efficient biofertilizers as per the quantity of nitrogen fixed concerned. They have seven genera and highly specific to form nodule in legumes, referred as cross inoculation group.

Azotobacter: They are the free living aerobic photo autotrophic, non-symbiotic bacteria.They secretes vitamin-B complex, gibberellins, naphthalene, acetic acid and other substances that inhibit certain root pathogens and improves root growth and uptake of plant nutrients. It occurs in the roots of Paspalumnotatum (tropical grasses) and other spp. andadds 15-93 kg N/ha/annum on *P. notatum* roots. *Azotobacterindícum* occurs in acidic soilin sugarcane plant roots. It can apply in cereals, millets, vegetables and flowers through sced, seedlings soil treatment.

Azospirillum: These are gram negative, free living, associative symbiotic and non-noduleforming, aerobic bacteria, occurs in the roots of dicots and monocot plants i.e. corn,sorghum, wheat etc. It is easy to culture and identify. *Azospirillum* is found to be very effective in increasing 10-15% yield of cereal crops and fixes N_2upto 20-40% kg/ha.Different A. brasiliense strains inoculation in the wheat seed causes increase in seed germination, plant growth, plumule and radicle length. *Herbaspirillum* species occurs inroots, stems and leaves of sugarcane and rice. They produce growth promoters (IAA. Gibberillins, Cytokinins) and enhance root development and uptake of plant nutrients (N,P& K).

Blue Green algae: Both free-living as well as symbotic cyanobacteria (blue greenalgae) have been harnessed in rice cultivation in India. A composite culture of BGA having heterocystous Nostoc, Anabaena, Aulosira etc. is given as primary inoculum in tray, polythene lines pots and later mass multiplied in the field for application as soil based flakes to the rice growing field at the rate of 10 kg/ha. The final product is not free fromextraneous contaminants and not very often monitored for checking the presence of desiredalgal flora.

Azolla: It is a free floating, aquatic fern found on water surface having ac yanobacteria symbiont Anabaena azollae in their leaves. It fixes atmospheric nitrogen andexcrete organic nitrogen in water during its growth and also immediately upon trampling. Azolla contributes nitrogen, phosphorus (15-20 kg/ha/month), potassium (20-25kg/ha/month) and organic carbon etc. and also suppresses weed growth. Azolla alsoabsorbs traces of potassium from irrigation water and can be used as green manure beforerice planting. *Azolla* spp. are mental tolerant hence, can be applied near heavy metalpolluted areas.

Phosphate solubilisingbacteria: *Pseudomonas fuorescens*, *Bacillus megatherium*var. *Phosphaticum*, *Acrobacteraerogens*, *Nitrobacter* spp., *Escherichia freundiil*, *Serratia* spp., *Pseudomonasstriata*, *Bacillus polymyxa* are the bacteria have phosphate solubilising ability. Phosphobacterin" are the bacterial fertilizers containing cells of bacillusmegatherium var. phosphaticum, prepared firstly by USSR scientists. They increased about10 to 20% crop yield and also produces plant growth promoting hormones which helps in phosphate solubilising activity of soil.

Plant Growth Promoting Rhizobacteria (PGPR): They are also called as microbialpesticides e.g. *Bacillus* spp. and Pseudomonas fluorescence. *Serratia* spp. and *Ochrobactrum* spp. are able to promote growth of plants. Pseudomonas fluorescence application to the Black pepper enhanced uptake of nutrients which increased plantbiomass. Fluorescent rhizobacteria improve the growth of *H. brasiliensis.*

Mycorrhiza: Mycorrhizas are developed due to the symbiosis between some specific root inhabiting fungi and plant roots and used as biofertilizers. They absorb nutrients suchas manganese, phosphorus, iron, sulphur, zinc etc. from the soil and pass it to the plant. Mycorrhizal fungus increases the yield of crops by 30-40% and also produces plant growth promoting substances.

Bio-compost: It is a kind of organic fertilizer, which is prepared from the waste of the sugar industry. The waste is decomposed using a number of human and plants friendly bacteria and Fungi. Bio-compost consists of nitrogen, phosphate solubilizing bacteria and plenty of useful fungi like the decomposing fungi. This bio-fertilizer helps the farmers to increase soil fertility and thereby increase the yield of the crops.

Vermi-Compost: It is also an organic fertilizer containing nitrogen phosphorus, potassium, sulphur, organic carbon, sulfur, hormones, enzymes, etc. If used over a period of time, the soil becomes extremely fertile and all the lost nutrients are restored back to the turf and the soil remains fertile. In this bioconversion process earthworms feed on the organic waste to produce vermicompost and vermiwash.

Trichoderma: It is an eco-friendly fertilizer, which acts as a bio-control agent and is hyper parasitic against different pathogens in the field.Trichoderma is a very effective biological mean for plant disease management especially the soil born. It is a free-living fungus which is common in soil and root ecosystems. It is highly interactive in root, soil and foliar environments. It reduces growth, survival or infections caused by pathogens by different mechanisms like competition, antibiosis, mycoparasitism, hyphal interactions, and enzyme secretion.Trichoderma strains solubilize phosphates and micronutrients. The application of Trichoderma strains with plants increases the number of deep roots, thereby increasing the plant's ability to resist drought.Trichoderma strains are known to induce resistance in plants. Three classes of compounds that are produced by Trichoderma and induce resistance in plants are now known. These compounds induce ethylene production, hypersensitive responses and other defense related reactions in plant cultivars. Trichoderma strains play an important role in the bioremediation of soil that are contaminated with pesticides and herbicides. They have the ability to degrade a wide range of insecticides: organochlorines, organophosphates and carbonates.

Method of Application

- **Seed treatment**: Mix 6-10 g of Trichoderma powder per Kg of seed before sowing.
- **Nursery treatment**: Apply 10-25 g of Trichoderma powder per 100 m2 of nursery bed. Application of neem cake and FYM before treatment increases the efficacy.
- **Cutting and seedling root dip**: Mix 10g of Trichoderma powder along with 100g of well rotten FYM per liter of water and dip the cuttings and seedlings for 10 minutes before planting.
- **Soil treatment**: Apply 5 Kg of Trichoderma powder per hector after turning of sun hemp or dhainch into the soil for green manuring. Or Mix 1kg of Trichoderma formulation in 100 kg of farmyard manure and cover it for 7 days with polythene. Sprinkle the heap with water intermittently. Turn the mixture in every 3-4 days interval and then broadcast in the field.
- **Plant Treatment**: Drench the soil near stem region with 10g Trichoderma powder mixed in a liter of water

Benefits: The advantages of Liquid Bio-fertilizer over conventional carrier based Bio-fertilizers are listed below:

- Longer shelf life - 12-24 months.
- No contamination.

- No loss of properties due to storage upto 45°C.
- Greater potentials to fight with native population.
- High populations can be maintained more than 109 cells/ml upto 12 months to 24 months
- Easy identification by typical fermented smell.
- Cost saving on carrier material, pulverization, neutralization, sterilization, packing and transport.
- Quality control protocols are easy and quick.
- Better survival on seeds and soil.
- No need of running Bio-fertilizer production units throughout the year.
- Very much easy to use by the farmer.
- Dosages is 10 time less than carrier based powder Bio-fertilizers.
- High commercial revenues.
- High export potential.
- Very high enzymatic activity since contamination is nil.

References

Ahemad M, Khan MS. Phosphate solubilizing Enterobacterasburiae strain PS2. Res J Microbiol. 2010. 5:849–857.

Bhatttacharyya, P. and Kumar, R. 2000. Liquid biofertilizer-current Knowledge and Future prospect. National seminar on development and use of biofertilizers, biopesticides and organic manures. Bidhan KrishiViswavidyalaya, Kalyani, West Bengal, November 10-12.

Hegde, S.V. 2008. Liquid bio-fertilizers in Indian agriculture. Bio-fertilizer news letter, pp.17-22.

LiinaSoonvald, KaireLoit, Eve Runno-Paurson, Alar Astover and LehoTedersoo, The role of long-term mineral and organic fertilisation treatment in changing pathogen and symbiont community composition in soil, Applied Soil Ecology, 10.1016/j.apsoil.2019.05.003, 141, (45-53), (2019).

Pierre Mariotte, Zia Mehrabi, T. MartijnBezemer, Gerlinde B. De Deyn, Andrew Kulmatiski, Barbara Drigo, G.F. (Ciska) Veen, Marcel G.A. van der Heijden and Paul Kardol, Plant–Soil Feedback: Bridging Natural and Agricultural Sciences, Trends in Ecology & Evolution, 10.1016/j.tree.2017.11.005, 33, 2, (129-142), (2018).

6

Celery Based Crop Rotations in Punjab Move towards Higher Returns and Farm Profitability

Sudeep Singh, Baljinder Kaur Sidana and Arjinder Kaur

Department of Economics and Sociology, Punjab Agricultural University Ludhiana, Punjab, India

Abstract

Specialty agriculture is the need of hour in a state like Punjab where scope for enhancing income in the prevailing cropping pattern is bleak with escalating costs, stagnant yields, nearly constant output prices leading to dwindling profit margins in the principal crops i.e. paddy and wheat. 'Celery' can be one such niche crop having potential for higher incomes to the debt ridden farming community of the Punjab state. The present study undertakes comparative economic analysis of celery based crop rotations using purposive sampling technique with a sample of 60 celery growers from 20 villages of two districts in the state. The gross returns are found to be 11 per cent higher in paddy-pea-celery rotation than paddy-pea-wheat rotation. Inspite of higher variable cost in paddy-pea-celery rotation, the incremental net returns (14.42 %) are found to be greater than incremental cost (8.13%) indicating profitability of including celery in the crop rotation. So farmers can shift sizeable area of wheat crop to celery in order to increase their farm income and promote crop diversification. Price fluctuations, delayed payment and weeds in the crop are some major constraints in celery cultivation, which needs to be addressed by creating awareness about usage of the crop in domestic pharmaceutical, hotel industry etc. IPM techniques can be encouraged in the management of disease and insect-pest attacks.

Keywords: Celery, Crop Rotation, Profitability

Introduction

Seasonality in agriculture is a major determinant of poverty, food insecurity and hunger in areas having harvest only once or twice a year. Undiversified livelihoods survive on the crops that the households consume as food and sell for income. The risk being high, a single failed harvest can cause huge losses and can destitute a marginalized household with limited or no savings, increasing their dependence on borrowings leading to indebtedness later on. Punjab being one of the pioneer states in green revolution in country is facing lot of challenges due to monoculture i.e. paddy-wheat as the major crop rotation followed on 92% area of the state. The situation has led to serious repercussions in agriculture sector of the state along with social, financial, and ecological issues. Speciality agriculture can be a breather in this grave situation and a move towards higher farm income for the distressed farming community of Punjab. It generally refers to high yield agriculture which is sustainable in various aspects. Speciality crops give rise to alternative crop rotations which increase income as well as cropping intensity.

Crop rotation is the practice of growing a series of different types of plants in the same area according to the growing season. It reduces pressure on a single set of nutrients, pesticides and weeds, and provide opportunities for integrated pest management technique. Numerous cropping systems can be possible on farmer's field. Decision criteria for the crop choice can include soil characteristics, environmental conditions and profitability from the crop. In addition, crop rotation can improve soil and biodiversity, reduce soil erosion and increase the stability of the farm system (Bullock, 1992).

The celery plant (Apium graveolens) is a hard biennial alimentary herb crop, belonging to the family Apiaceae. It is a medicinal plant cultivated for its fleshy leafstalk used in spicing up foods and drinks and used as a vegetable along with essential oil extracted from the seeds (Tyagi, 2013). The main cultivated regions of celery are Asia, America and Europe. In America, the major states where celery is cultivated are Florida, New York, California and Michigan. It is cultivated in European nations like Germany, Hungary, Italy, Holland, United Kingdom, France and Belgium. In Western countries, celery is one of the most important crop being cultivated after lettuce, mainly consumed as a salad but also be cooked as vegetables. In the Asian region like China and India, it is cultivated for its seed which is used as spice. (Malhotra, 2006).

The origin of the celery is the Mediterranean region. In the sixteenth century, celery was cultivated in France and England as a medicinal plant, vegetable crop and seed crop which yield essential oil. Later, it spreads to many part of temperate Europe and USA. In 1930, the crop was introduced in India from France by a trading company in Amritsar (Punjab) and now commercially grown as a seed crop. In India, Punjab (Amritsar, Tarn Taran, Gurdaspur, Patiala) Haryana

(Panipat) and Uttar Pradesh (Saharanpur) are leading states in celery cultivation (Vijay and Malhotra, 2002).

India is a major producer and exporter of celery seeds in the world market that contributes to oil and oleoresin extraction. Celery worth Rs. 5310 lakhs was exported in 2015-16 which increased to Rs. 6649 lakhs in 2018-19 (Spice Board, 2018-19). The high yielding varieties recommended for cultivation of celery in India includes EC-99249-1 and PRL-85-1 and give average yield of 20 quintals per hectare (NHB, 2019).

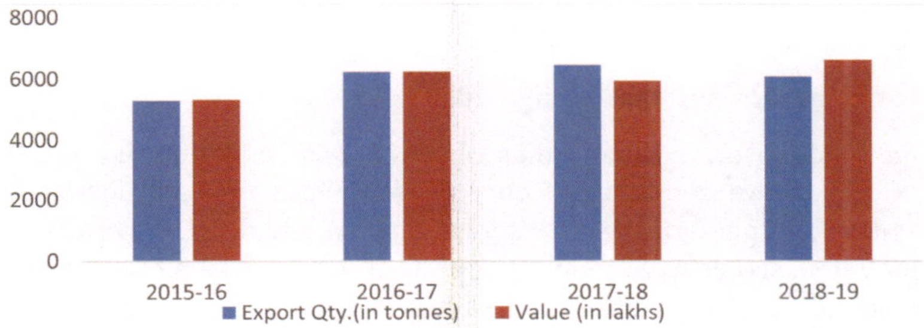

Fig. 1: Export of Celery from India, 2015-16 to 2018-19
Source: Spice Board, Ministry of Commerce and Industry, GOI, 2018-19

Punjab is one of the major celery producing states of India, contributing about 90% to the total production in the country (NHB, 2019). The present study have been specially designed to examine the profitability of different crop rotations including this special crop 'celery' being cultivated in vegetable pockets of the state.

Data and Methodology

Primary data from 60 celery growers was collected from two districts namely Amritsar and Patiala covering 20 villages for the year 2020-21. A multi-stage purposive sampling technique was adopted for the selection of the sample. At the first stage two districts i.e. Amritsar and Patiala have been selected keeping in view the major proportion of total cultivated area of celery in these districts. As the celery growers were scattered over many villages, thus 14 villages from Amritsar district and six from Patiala district became the part of the present study. The primary data were collected from the sample farmers by filling the information in detailed schedules through personal interview method. The information regarding the operational holding, cropping pattern, area under celery cultivation, detailed cost on celery cultivation, and their related crops involved in rotation, year of cultivation, input used, production, price were obtained from the farmers.

Descriptive statistics and enterprise budgeting was used for analysing the results.

2.1 Garrett Ranking Technique

Garrett ranking method was used to evaluate the problems faced by sample farmers in the marketing celery. The farmers were asked to rank the given problem according to the magnitude of the problem. The orders of merit given by the respondents were converted into ranks by using the following formula:

$$\text{Percent position} = \frac{100\left(R_{ij} - 0.5\right)}{N_j}$$

Where,

R_{ij} = Rank given for the ith item by the jth individual

N_j = Number of items ranked by the jth individual

The percentage position of each rank thus obtained was converted into scores by referring to the table provided by Henry Garrett. Then for each factor the scores of individual sample farmers were added together and divided by the total respondents for whom scores were added. Thus, mean score for each problem was ranked by arranging them in the descending order.

Results and Discussion

3.1 Trends in Celery Production in Punjab

'Karnauli' is the common name for the celery in Punjab. The variety which is recommended by Punjab Agricultural University, Ludhiana is 'Punjab Celery 1'. Recently ITC has taken an initiative to purchase the produce from the celery growing farmers directly from their fields or transport the produce to their nearby godowns thus creating a permanent market for the crop. This crop has export value and therefore, expected that if farmers reduce some area under wheat and replace it with celery, they can earn more money, if grown successfully.

Area under celery in Punjab in year 2018-19 was 4496 ha out of which highest area i.e.1766 hectare was in Amritsar with total production of 2412 metric tonnes and productivity of 1366 kg per hectare (GOP, 2017-18). Figure 2 clearly depicted the increasing trend in area, production and yield of celery from 2013-14 to 2018-19 in Punjab state.

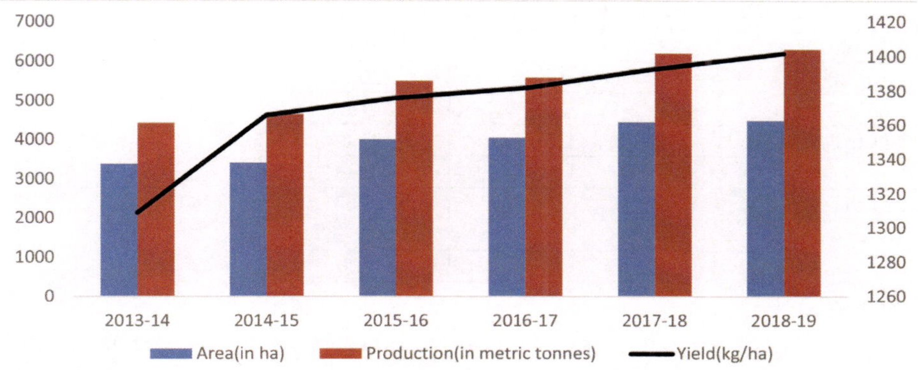

Fig. 2: Area, production and yield of celery in Punjab
Source: Department of Horticulture, GOP, 2017-18

3.2 Socio-Economic Profile of Sample farmers

Certain characteristics of the individual like education, age, family size etc. have an impact on rational decision making about area under crop. So, some attributes of celery growers have been discussed in Table 1. Majority of celery growers belong to middle age group with experience of more than 30 years in celery cultivation and acquired education upto senior secondary level. The average operational holding with the sample farmers was 9.60 hectares(ha.), out of which only 2.64 ha. were under celery crop, which was about 27.5 per cent of the average operational holding.

Table 1. Descriptive statistics of sample farmers, Punjab

Variable	Mean	Std. Dev.	Min	Max
Age(yrs)	44.13	12.82	24.00	85.00
Schooling(years)	13.55	3.31	0.00	19.00
Family size(no.)	6.63	2.59	3.00	16.00
Celery Area(ha.)	2.64	4.87	1.00	30.00
Operational Area(ha.)	9.60	16.50	4.00	100.00

3.3 Cropping Intensity and Area under Celery Crop

Majority of farmers (42%) cultivated about 1.20 to 2.40 ha of land under celery cultivation; whereas only three percent of farmers brought more than 6 ha under celery crop. This was due to price and production risk associated with this crop and high labour requirement. About 17 per cent of sample farmers went for more than 3.60 ha of land under this crop. The medium and large sized farmers well versed with agronomic practices of this crop were putting on an average 4.80 ha of land under this crop.

Table 2. Area under celery crop among sample farmers, Punjab, 2020-21

Area of celery	Number	Percentage
0 to 1.20	12	20.00
1.21 to 2.40	25	41.67
2.41 to 3.60	13	21.66
3.61 to 4.80	5	8.33
4.81 to 6.00	3	5.00
Above 6.00	2	3.33
Total sample	60	100

(hectares)

The cropping pattern among sample farmers indicates that average net sown area was 9.60 ha of the sample farmers. Paddy and wheat dominated in kharif and rabi season which accounted for 35.48 per cent and 23.14 per cent of gross cropped area respectively (Annexure 1). Fodder was grown in kharif season after paddy with percentage share of 1.75 per cent in gross cropped area. In pre-rabi and rabi season, peas and celery also covered a sizeable portion of gross cropped area grown on about 16.94 per cent and 10.46 per cent respectively, Potato crop also covered almost 5.30 per cent of the gross cropped area.

3.4 Different Crop Rotations Followed by Sample Farmers

Celery (rabi crop) needs warm days and cool nights, low humidity and plenty of sunshine. The cool temperature at the time of sowing and dry weather at seed ripening stage is ideal for this crop. The crop grows well on loamy soil. The seedlings are raised in nursery-beds and 60 to 70 days old seedlings are transplanted in the fields. The most important feature of this crop is that if it is grown on same piece of land again, then it germinates naturally from the shattered seeds of previous year and requires no tillage. The sample farmers have not raised seedlings rather they went for broadcasting of seed or the crop germinated itself from shattered seeds of previous year, thus on an average 930 grams of seed per acre was used by celery growers.

The major celery based rotations adopted by sample farmers were paddy-pea-celery, paddy-beans-celery, paddy-potato-celery and paddy-cabbage-celery. Among celery growers, paddy-pea-celery was the most preferred crop rotation which was followed by 98 per cent of sample farmers because according to them, pea fits best with celery and after 2-3 pickings give adequate time to celery for its growth and development. The pea crop is uprooted after 2-3 pickings and then hoeing is given for celery crop for proper spacing and germination. Proper plant population is maintained by thinning wherever needed. Although celery can be cultivated after bean, cabbage and potato crops but in case of potato, after harvesting main crop, celery seedlings have to be transplanted in the field which increases the cost of production. This was evident from the table that only 5 to 10 per cent of the sample farmers cultivated pre rabi crop other than pea before celery crop. All the sample farmers allocated major proportion of rabi area to wheat crop (about 25%) and 10-15 per cent of their gross cropped area to celery (after pea) crop. As such, the sample farmers cultivated both crops in rabi season i.e. wheat as well as celery. About 52 per cent of sample farmers followed paddy-pea-wheat rotation.

Table 3. Crop rotations adopted by sample farmers in Punjab, 2020-21 (Multiple response)

Crop Rotation	Number of farmers	Percentage
Paddy-pea-celery	59	98.30
Paddy-pea-wheat	31	51.66
Paddy-bean-celery	3	5.00
Paddy-cabbage-celery	6	10.00
Paddy-potato-celery	3	5.00

3.5 *Cost and Return Structure of Different Crops as well as Crop Rotations Followed by Sample Farmers*

The cost and return structure of different crops cultivated by sample farmers have been presented in the Table 4. Due to the highest productivity, higher per hectare gross returns for potato was realized as Rs. 208886 followed by cabbage (Rs. 150389) and beans (Rs. 138473). The per hectare variable cost on potato crop was Rs. 115205 due to high cost of seed and hired labour involved in harvesting. Wheat incurred lowest variable cost among the different crops cultivated by sample farmers. Returns over variable cost of celery when compared with wheat depicted that 50 per cent higher returns were obtained from celery but 19.2 percent lower returns were obtained from celery when compared with potato crop. Output-Input ratio for all the crops was greater than one because gross returns obtained from the crop were higher than cost of cultivating crops and highest output-input ratio of 2.84 was realized for paddy followed by celery (2.79) and wheat (2.69), which showed that a farmer earned Rs. 2.84, 2.79 and 2.69 by investing rupee one on paddy, celery and wheat crops respectively.

Table 4. Cost and return structure of different crops cultivated by sample farmers, Punjab, 2020-21.

Particulars	Celery	Pea	Bean	Cabbage	Wheat	Paddy	Potato
Average yield (quintals)	14.10	76.75	74.85	232.80	40.50	70.50	271.28
Average price (Rs./Qtl)	8350	1642	1850	646	1975	1888	770
Gross returns (Rs./ha)	117735	126024	138473	150389	79986	133104	208886
Variable cost (Rs./ha)	42070	75915	65995	92475	29680	46835	115205
Returns over variable cost (Rs./ha)	75665	50109	72478	57914	50306	86269	93681
Output-input ratio	2.79	1.66	2.09	1.62	2.69	2.84	1.81

A perusal of the table 5 revealed that per hectare net returns of Rs. 212045 were estimated from paddy-pea-celery which was around 17 per cent less when compared with paddy-potato-celery. Paddy-pea-celery gave about 3 to 17 per cent less returns when compared with other rotations. This was due to availability of short duration varieties of pea which were harvested within 60-70 days of sowing, thus maintaining proper time for growth of celery crop. Hence, the majority despite of less returns than paddy-potato-celery went for paddy-pea-celery rotation.

Table 5. Cost and return structure of celery based crop rotation by sample farmers in Punjab, 2020-21 (Rupees per ha)

Crop Rotation	Variable cost	Gross return	ROVC
Paddy-pea-celery	164820	376865	212045
Paddy-cabbage-celery	181380	401230	219850
Paddy-bean-celery	154900	389313	234413
Paddy-potato-celery	204110	459723	255613

3.6 Comparative Returns from Paddy-Pea-Celery Vs Paddy-Pea-Wheat Crop Rotation

Celery, being a winter season crop competes with other rabi season growing crops, as such a comparison has been drawn with two major crop rotations paddy-pea-wheat and paddy-pea-celery. The perusal of table 6 showed that gross returns were estimated to be 11.13 per cent higher from paddy-pea-wheat than paddy-pea-celery. Although variable cost in paddy-pea-celery rotation was higher but per cent increment in net returns (13.59%) was more than per cent increment in cost (8.13%) indicating the profitability by including celery in crop rotation. Farmers can shift some wheat area to celery in order to increase their farm income and promote crop diversification.

Table 6. Comparative returns from paddy-pea-celery vs paddy-pea-wheat crop rotation in Punjab, 2020-21 (per ha)

Particulars	Paddy-pea-celery (1)	Paddy-pea-wheat (2)	Percent change* over (2)
Gross returns	376865	339118	11.13
Total variable cost	164820	152430	8.13
Returns over variable cost	212045	186688	13.59

*Percent change signifies increase in returns from paddy-pea-wheat rotation

3.7 Problems in the Marketing of Celery

Profitability in cultivation is not the only factor tempting the farmers to increase area under any crop. Marketing of produce is another major aspect involved in rational decision making. Ultimately the remunerative price realized of output in the market will lead to profits. In this light certain constraints faced by celery growers in marketing of the crop have been discussed in table 7. Among various marketing problems faced by the sample farmers, price fluctuation for crop produce was the major problem with 65.08 Garrett's score. The second and the third rank was given to delayed payment for the crop and inadequate facility of unloading the crop by the sample farmers. Malfunctioning by dealers and high transportation cost were other marketing problems reported by farmers.

Table 7. Problems faced by the sample farmers in the marketing of celery, Punjab, 2020-21

Problems	Per cent position	Garrett's Value	Average Score	Rank
Price fluctuation	10	75	65.08	I
Delayed payment for crop	30	60	52.95	II
Inadequate facility of unloading the crop	50	50	51.18	III
Malfunctioning by Dealers	70	40	40.96	IV
High transportation cost	90	24	32.66	V

3.8 Suggestions Given by the Farmers

Suggestions were sought from sample farmers to remove the bottlenecks in marketing of celery crop and ease the business. This crop is highly labour intensive and as such more than half of the total expenditure i.e. 68.29 per cent was made on payment to hired labour. Due to this, it was suggested by all sample farmers for the manufacturing of proper machinery to replace labour(Table 8). With average score of 55.48, the sample farmers suggested that short duration varieties should be developed for this crop. The sample farmers demanded more price of the crop against their produce and suggested to develop disease resistance varieties which occupied third and fourth rank, respectively. The farmers were not much aware about insect-pest control and proper harvesting procedures, as such, regular and more effective extension about the crop was suggested by the farmers which occupied fifth rank.

Table 8. Suggestions given by the sample farmers to overcome the problems faced in farming of celery, Punjab, 2020-21

Suggestions	Per cent position	Garrett's Value	Average score	Rank
Machinery to replace labour	10	75	59.83	I
Develop short duration varieties	30	60	55.48	II
More prices	50	50	51.23	III
Develop disease resistant varieties	70	40	50.25	IV
Regular and more effective extension	90	24	42.65	V

Conclusion and Policy Implications

The celery cultivation was mostly adopted by large and medium farmers. Among celery growers, paddy-pea-celery was the most preferred crop rotation which was followed by 98 per cent of sample farmers but this rotation gave about 3 to 17 per cent less returns when compared with other celery based rotations. The important constraints in the cultivation of this crop were high intensity of weeds, high labour needs, erratic electricity supply and price variability whereas price fluctuation, delayed payment, inadequate facilities of unloading the crop and malfunctioning by dealers were the major marketing problems faced by the farmers.

Domestic demand of celery in pharmaceutical is less than 10 per cent and this leads to price volatility in domestic market. There is huge potential to increase the industrial use of celery especially in pharmaceutical, flavouring and hotel industry. In addition diverse food products can be developed using celery. So farmers can shift sizeable area of wheat crop to celery in order to increase their farm income and promote crop diversification. IPM techniques should be used in the management of insect-pest and diseases. Instead of chemical pesticide, neem based bio-pesticides should be used to manage the insect pest at its initial stages of growth.

References

AGMRC (2018) Agricultural Marketing Resource Centre/ Celery (https://www.agmrc.org/commodities-products/vegetables/celery)

Bullock D. G. (1992) Crop rotation. Critical Reviews in Plant Sciences. Vol 11, No.4, pp: 309-26.

GOP(2017-18) Celery Statistics, Department of Horticulture, Government of Punjab, Punjab

Malhotra S.K. (2006) Handbook of Herbs and Spices. pp. 317-36 United kingdom Elsevier Science, United Kingdom. https://www.researchgate.net/publication/323691943_18_Celery

NHB (2019) A Model Bankable Project On Celery. National Horticulture Board, Ministry of Agriculture, Government of India, Gurgaon, Haryana. http://nhb.gov.in/modelproject-reports/Horticulture%20Crops/Celery/Celery1.htm).

Spice Board (2018-19) Celery export, Ministry of Commerce and Industry, Government of India.

Tyagi S., Chirag P., Mangukia D., Manjukia D., Gupta A. K. (2013) Medicinal Benefits of Apium Graveolens (Celery herb). J Drug Discov Ther Vol 1,No.5 : 36-38.

Vijay O. P. and Malhotra S. K. (2002) Seed spice in India and World. Seed Spices Newsletter, Vol 2, No. 1, pp: 1-4

Annexure. Cropping pattern among sample farmers, Punjab, 2020-21

Crop	Area (acres)	Percentage
Kharif		
Paddy	8.96	35.48
Sugarcane	0.20	0.79
Kharif fodder	0.45	1.75
Pre- Rabi		
Peas	4.28	16.94
Potato	1.33	5.30
Beans	0.39	1.55
Cabbage	0.08	0.33
Cauliflower	0.06	0.24
Rabi		
Wheat	5.85	23.14
Celery	2.64	10.46
Winter maize	0.43	1.68
Rabi Fodder	0.36	1.41
Tomato	0.13	0.52
Cucumber	0.04	0.19
Chicory	0.03	0.13
Sugarbeet	0.02	0.09
Gross Cropped Area	25.25	100.00
Cropping Intensity (%)	263	

7

Profitability and Marketing Efficiency of Rabi Pea in Different Agro-Climatic Zones of Himachal Pradesh

Parul Barwal , Subhash Sharma and Diksha Bali

Department of Social Sciences, Dr. Yashwant Singh Parmar University of Horticulture and Forestry Nauni, Solan, Himachal Pradesh, India

Abstract

The present search was undertaken to analyze the cost of cultivation and returns of mango cultivation in Kangra district of Himachal Pradesh.. A multistage sampling technique was employed for selecting 100 mango growers. The cost of cultivation details revealed that mango cultivation was highly a capital intensive crop. Maintenance cost of bearing mango per hundred plants at overall level showed an increase in cost up to main bearing stage and thereafter, it started decreasing which is due to the decrease in production and lead to lesser labor requirement. On an average, the total cost of cultivation per hectare for main bearing stage was found to be Rs. 57861.69in which the human labor component alone accounted for 37.44 per cent. Further, average production varied from 50.96 quintals to 53.00 quintals per 100 plants and net returns varied from Rs. 178365.54 to Rs. 185507.41 in different farm categories. Overall, the study recommends that the mango growers should improve the adoption of existing package-of-practices along with proper utilization of resources in order to cut down the cultivation cost towards realizing higher net returns.

Keywords: Mango, Cost-Returns, Profitability, Production

Introduction

Agriculture is the backbone of Indian economy. There are a number of sub sectors in the agricultural sector such as sericulture, floriculture and horticulture. Among those fields horticulture has played a dominant role in the total agricultural output throughout India. The world market has provided a good opportunity for the development of the fresh and the processed fruits industry of India. From the point of view of domestic market also, horticulture products have provided food items of nutritional value. Their cultivation has generated higher employment opportunities and higher returns to the farmers than the cereals and other agricultural cash crops (Singhal, 1995). Mango (*Mangifera indica*) is a stone fruit belonging to family Anacardiaceae. It is a most important fruit of India and known as the "king of tropical fruits". Mango is cultivated in many tropical regions and distributed widely in the world. Nearly 50 different varieties of mango are cultivated in India. Alphonso and Kesar mango varieties are the best mango varieties in southern states of India, while Dassehari and Langda varieties are most popular in the northern states. Raw fruits of local varieties of mango trees are used for preparing various traditional products like raw slices in brine, amchur, pickle, murabba, chutney, etc. World demand for mango is increasing, particularly from temperate countries where mangoes are rapidly gaining popularity. Leading mango importing countries are USA, France, England, Portugal, Spain, Belgium, Denmark, and Sweden. Apart from India, other major producers of mango are China, Mexico, Thailand, Indonesia, Pakistan, Philippines, Nigeria, Brazil, Peru, Australia, South Africa, Malaysia and Venezuela. It is a national fruit of India, Pakistan and Philippines. It is cultivated in an area of 2,209 thousand hectares with a production of 18,643 million tonnes in India and contributes 40 per cent of total world production of mango (NHB, 2016).

Himachal Pradesh is having the advantage of varied climate, ranging from subtropical to dry temperate. The total fruit production in the state is 9,28,829 metric tonnes from an area of 2,26,799 hectares. Production of mango is distinguished from annual crops by the long gestation period, an extended period of output flow and varying stages of productivity over the lifetime (Chand, 1994). Therefore, it is more difficult to determine economics of growing mango as compared to the annual crops. Having sufficient awareness regarding profitability of any enterprise is needed to make rational decisions at the farm level. The producers would like to know the results of this economic activity by working out detailed costs and benefits of the investment in the project (Akcay & Uzunoz, 2005). With this background, the study was conducted to evaluate the cost of cultivation of mango in Kangra district of Himachal Pradesh.

Sampling Procedure

Selection of the study area

Kangra district of Himachal Pradesh was purposively selected for the present study because of its significant contribution in area and production of mango in Himachal Pradesh. The area under mango in Kangra district is 21,286 hectares which is 51.26 per cent of total area under mango cultivation of H.P and production is 24,900 metric tonnes i.e. 47.70 per cent of total production of Himachal Pradesh (Directorate of Economics and Statistics, 2016).

Selection of the sample

Multistage Random Sampling technique was used to select the ultimate respondents i.e. the mango growers. At first stage, two blocks of Kangra District (Nurpur and Indora) were selected on the basis of area under mango cultivation. At the second stage, a list of the villages in the selected blocks was prepared and 8 villages from each selected blocks were selected. At the third stage, a list of farmers having mango orchards of the selected villages was prepared with the help of the officials from the Revenue and Horticulture departments and 5 mango growers were selected from each village, forming a sample of 80 farmers for the present study.

Nature and Sources of Data

To meet the objectives of the present study, both primary as well as secondary data were collected.

Primary data

The Primary data were collected from the sampled growers and market functionaries by the survey method using well designed and pretested schedule during the year 2017-18. The data collected from the growers comprised data on demographic features such as family size, age, education, economic parameters (land use pattern, livestock, cropping pattern, irrigation status and income), cost of production, yield, marketing costs, marketing margins, various charges in market and problem faced by growers in various aspects of production and marketing from the selected households in the study area.

Secondary data

Secondary data pertaining to the list of village households, area, production, and productivity of mango at district and state levels were collected from the Directorate of Horticulture, Shimla, National Horticulture Board, Gurugram, Block Development offices of respective blocks.

Analytical Framework

To fulfill the specific objectives of the study, based on the nature and extent of availability of data, the costs and returns structure of mango was analysis

Costs and returns

In determining the costs and returns of perennial crops, age of plantation plays an important role. Keeping in view this point, the sample has been further classified into different groups according to age of mango plants. First group consisting of the year of plantation considered as zero year showing the initial cost of plantation. Second group consisted of non-bearing plants from 1-5 years considered as non bearing stage and third group consisted of bearing stage which has been further divided into three sub groups i.e. early bearing (6-15 years), main bearing (16-35 years) and late bearing (>35 years).

The total age of mango plantation was considered to be 60 years. Following cost concepts were used.

Plantation cost: All costs incurred by mango growers from preparation of land upto planting of plants.

(i)	Plantation cost	=	Land preparation (pit digging + layout) + Cost of plants and other inputs.
(ii)	Fixed cost	=	Land revenue + Rental value of land + Depreciation + Interest on fixed capital
(iii)	Variable cost	=	Cost of inputs used including labour + interest on working capital
(iv)	Establishment cost	=	Plantation cost + cost during non-bearing stage.
(v)	Maintenance cost	=	Establishment cost and cost during bearing stage

Results

Costs and Returns of Mango Cultivation

The cost of cultivation is a subject of attention to a wide range of users of cost data and it assumes particular importance in the area of planning. The cost of cultivation of horticultural commodities for planning is appreciated as these data inform the planners about the area where it is economical to produce various commodities and the regions that would accordingly be most appropriate for the development of industries based on the agricultural raw material. These facilitates the experts to make practical recommendations for farm planning aimed at better allocation of existing resources and introduction of improved agronomic practices that would increase the competence in mango production.

The summation of cost incurred during plantation (initial investment) and costs during non-bearing stage (establishment cost) which is distributed over total productive life of the plantation.

The costs and returns were worked out on hundred plant basis under marginal, small, medium and overall farm categories. For estimating the costs and returns from mango, it has been assumed that:

i) First bearing starts from 6th year in the study area.

ii) The major operations and inputs requirement remains same in early bearing stage (6- 15 year), main bearing stage (16-35 year), and late bearing stage (> 35 year) .

iii) Total economic life of the plantation is 60 years in the study area.

The above mentioned stages are based on growth and productivity pattern of the plant. The analysis in the section has been divided into five sub-heads.

1.1 Cost of plantation (initial investment) in different farm categories.

1.2 Cost during non-bearing stage of mango plantation in different farm categories.

1.3 Cost during bearing stage of mango plantation in different farm categories.

1.4 Returns from mango cultivation in different farm categories.

In mango, initial investment is quite heavy for reasons of the cost involved in land development, digging of pits, application of manure & fertilizers and material cost. Growers have to incur cost on maintenance for about 5th years without any returns.

1.1 Cost of plantation (Initial investment)

Cost of plantation consists of all the expenditures incurred on different inputs and labour required for the establishment of an orchard.

Table 1: Initial costs of mango orchard in the study area

(Rupees/100 Plants)

Sr.No.	Particulars	Initial investment			
		Marginal	Small	Medium	Overall
1	**Material Cost**				
a)	Planting material	8235.29	8250.00	8312.50	8259.97
		(25.60)	(23.88)	(21.98)	(23.64)
b)	FYM	1372.55	1416.67	1600.00	1445.49
		(4.27)	(4.10)	(4.23)	(4.14)
c)	Soil treatment	274.51	279.17	287.50	279.32
		(0.85)	(0.81)	(0.76)	(0.80)
	Total material cost	9882.35	9945.83	10200.00	9984.78
		(30.72)	(28.79)	(26.97)	(28.58)
2	**Human labour**				
a)	Family labour	6382.35	7218.75	6037.50	6542.75
		(19.84)	(20.89)	(15.96)	(18.73)

Sr.No.	Particulars	Initial investment			
		Marginal	Small	Medium	Overall
b)	Hired labour	0.00	583.33	3106.25	990.39
		(0.00)	(1.69)	(8.21)	(2.83)
	Total human labour	6382.35	7802.08	9143.75	7533.14
		(19.84)	(22.58)	(24.18)	(21.56)
3	Interest on working capital	494.12	526.46	665.31	548.76
		(1.54)	(1.52)	(1.76)	(1.57)
4	Risk Margin	988.24	1052.92	1330.63	1129.63
		(3.07)	(3.05)	(3.52)	(3.23)
5	Managerial cost	988.24	1052.92	1330.63	1129.63
		(3.07)	(3.05)	(3.52)	(3.23)
6	Total Variable Cost	18735.29	20380.21	22670.31	20660.86
		(58.24)	(58.99)	(59.94)	(59.13)
7	**Fixed cost**				
a)	Land revenue	31.00	31.00	31.00	31.00
		(0.10)	(0.09)	(0.08)	(0.09)
b)	Deprecation	1180.31	1719.48	2399.15	1662.01
		(3.67)	(4.98)	(6.34)	(4.76)
c)	Interest on Fixed capital	446.68	644.40	944.93	636.79
		(1.39)	(1.87)	(2.50)	(1.82)
d)	Rental value of land	11775.00	11775.00	11775.00	11775.00
		(36.60)	(34.08)	(31.13)	(33.70)
	Total fixed cost	13432.99	14169.88	15150.07	14279.60
		(41.76)	(41.01)	(40.06)	(40.87)
8	Total cost	32168.29	34550.09	37820.39	34940.45
		(100.00)	(100.00)	(100.00)	(100.00)

Figure in parentheses are percentage to total cost

It comprises of specifically expenditure on land preparation and digging of pits along with FYM and planting cost. It also includes cost on human labour (hired and family), land revenue, risk margin and managerial cost.

The item wise plantation cost incurred in the initial year has been presented in Table 1. It is clear from the Table 4.18 that, mango orchardists in study area incurred on an average, total cost of Rs.34940.45 per hundred plants. The per cent share of planting material to the total cost was maximum (23.64%) followed by FYM (4.14%) and soil treatment (0.80%). Human labour contributing 21.56 per cent to the total cost at overall level which varied from 19.84 per cent in marginal farm to 24.18 per cent in medium farm category. Per cent share of interest on working capital was 1.75 per cent and risk margin & managerial cost contributed equally to the total cost was i.e 3.23 per cent at an overall level. Total variable cost varied from 58.24 per cent in marginal farm to 59.94 per cent in medium farm and at overall level was 59.13 per cent. Total fixed cost was 41.76 per cent on marginal farm , 41.01 per cent on small farm and 40.06 per cent on medium farm category respectively. Rental value of land shared 33.70 per cent of the total cost at overall level. Total cost varied from Rs. 32168.29 to Rs. 37820.39 among different farm categories.

1.2 Cost during non-bearing Stage of Mango Plantation in Different Farm Categories

The item wise maintenance cost during non-bearing stage of mango for marginal, small, medium and overall farms were presented in Tables 2 to Table 5. The total maintenance cost of non- bearing mango plantation has shown positive relationship with age of the plants and was estimated to Rs. 38938.04, Rs. 40965.02, Rs. 43982.91 Rs. 46939.33 and Rs. 50363.15 per hundred plants in the first year to fifth year respectively (Table 2). The per cent share of total material cost varied from 4.41 to 4.72 per cent from first to fifth year. Total labour cost was 42.24 per cent in the fifth years to 54.76 per cent in the first year. Marginal farmers used all the family labour for different operations. Total variable cost ranged between 48.14 to 60.34 per cent from fifth to first year and fixed cost varied from 39.66 per cent to 51.86 per cent from first to fifth year. Risk margin and managerial cost exhibited positive trends with the age of plantation. The per cent share of depreciation cost and rental value of land in the total cost declined with the increase in the age of plants; however, absolute value remains the same.

Table 2. Maintenance cost during non- bearing stage of mango orchard in case of marginal farms

(Rupees/100 Plants)

Sr. No.	Particulars	Age of plantation in years				
		1st year	2nd year	3rd year	4th year	5th year
1	**Material Cost**					
a)	Gap filling	83.33	-	-	-	-
		(0.21)				
b)	Bordeaux mixture	125.00	181.82	212.12	235.29	555.56
		(0.32)	(0.44)	(0.48)	(0.50)	(1.10)
c)	FYM	1300.00	1333.33	1454.55	1529.41	1500.00
		(3.34)	(3.25)	(3.31)	(3.26)	(2.98)
d)	Plant protection chemical (0.59)	230.63	290.91	301.82	305.88	321.11
		(0.71)	(0.69)	(0.65)	(0.64)	
	Total material cost	1738.96	1806.06	1968.48	2070.59	2376.67
		(4.47)	(4.41)	(4.48)	(4.41)	(4.72)
2	**Human labour**					
a)	Family labour	21320.83	20830.30	21084.85	21164.71	21272.22
		(54.76)	(50.85)	(47.94)	(45.09)	(42.24)

Sr. No.	Particulars	Age of plantation in years				
		1st year	2nd year	3rd year	4th year	5th year
	Total human labour	21320.83	20830.30	21084.85	21164.71	21272.22
		(54.76)	(50.85)	(47.94)	(45.09)	(42.24)
3	Interest on working capital (0.22)	86.95	90.30	98.42	103.53	118.83
		(0.22)	(0.22)	(0.22)	(0.24)	
4	Risk Margin	173.90	180.61	196.85	207.06	237.67
		(0.45)	(0.44)	(0.45)	(0.44)	(0.47)
5	Managerial cost	173.90	180.61	196.85	207.06	237.67
		(0.45)	(0.44)	(0.45)	(0.44)	(0.47)
6	Total Variable Cost	23494.53	23087.88	23545.45	23752.94	24243.06
		(60.34)	(56.36)	(53.53)	(50.60)	(48.14)
7	**Fixed cost**					
a)	Land revenue	31.00	31.00	31.00	31.00	31.00
		(0.08)	(0.08)	(0.07)	(0.07)	(0.06)
b)	Deprecation	1180.31	1180.31	1180.31	1180.31	1180.31
		(3.03)	(2.88)	(2.68)	(2.51)	(2.34)
c)	Interest on Fixed capital	446.68	446.68	446.68	446.68	446.68
		(1.15)	(1.09)	(1.02)	(0.95)	(0.89)
d)	Rental value of land	11775.00	11775.00	11775.00	11775.00	11775.00
		(30.24)	(28.74)	(26.77)	(25.09)	(23.38)
e)	Interest on past capital	2010.52	4444.15	7004.46	9753.39	12687.10
		(5.16)	(10.85)	(15.93)	(20.78)	(25.19)
	Total fixed cost	15443.51	17877.14	20437.45	23186.38	26120.09
		(39.66)	(43.64)	(46.47)	(49.40)	(51.86)
8	Total cost	38938.04	40965.02	43982.91	46939.33	50363.15
		(100.00)	(100.00)	(100.00)	(100.00)	(100.00)

Figure in parentheses are percentage to total cost

Table 3. Maintenance cost during non- bearing stage of mango orchard in case of small farms (Rupees/100 Plants)

Sr.No.	Particulars	Age of plantation in years				
		1st year	2nd year	3rd year	4th year	5th year
1	Material Cost					
a)	Gap filling	220.99	-	-	-	-
		(0.58)				
b)	Bordeaux mixture	234.81	277.78	291.67	314.21	316.46
		(0.61)	(0.68)	(0.66)	(0.66)	(0.62)
c)	FYM	1685.08	1722.22	1755.56	1803.28	1792.41
		(4.41)	(4.19)	(3.98)	(3.78)	(3.54)
d)	Plant protection chemical	331.49	350.22	361.39	377.05	379.75
		(0.87)	(0.85)	(0.82)	(0.79)	(0.75)

Sr.No.	Particulars	Age of plantation in years				
		1st year	2nd year	3rd year	4th year	5th year
	Total material cost	2472.38	2350.22	2408.61	2494.54	2488.61
		(6.46)	(5.71)	(5.46)	(5.23)	(4.91)
2	**Human labour**					
a)	Family labour	17377.53	17908.33	18167.59	18398.91	18111.39
		(45.43)	(43.52)	(41.20)	(38.56)	(35.73)
b)	Hired labour	1160.22	1263.89	1302.78	1721.31	1949.37
		(3.03)	(3.07)	(2.95)	(3.61)	(3.85)
2	Total human labour	18537.75	19172.22	19470.37	20120.22	20060.76
		(48.47)	(46.60)	(44.15)	(42.17)	(39.58)
3	Interest on working capital	181.63	180.71	185.57	210.79	221.90
		(0.49)	(0.44)	(0.42)	(0.44)	(0.44)
4	Risk Margin	363.26	361.41	371.14	421.58	443.80
		(0.95)	(0.88)	(0.84)	(0.88)	(0.88)
5	Managerial cost	363.26	361.41	371.14	421.58	443.80
		(0.95)	(0.88)	(0.84)	(0.88)	(0.88)
6	Total Variable Cost	21918.28	22425.97	22806.83	23668.72	23658.86
		(57.31)	(54.50)	(51.72)	(49.60)	(46.67)
7	Fixed cost					
a)	Land revenue	31.00	31.00	31.00	31.00	31.00
		(0.08)	(0.08)	(0.07)	(0.06)	(0.06)
b)	Deprecation	1719.48	1719.48	1719.48	1719.48	1719.48
		(4.50)	(4.18)	(3.90)	(3.60)	(3.39)
c)	Interest on Fixed capital	644.40	644.40	644.40	644.40	644.40
		(1.68)	(1.57)	(1.46)	(1.35)	(1.27)
d)	Rental value of land .	11775.00	11775.00	11775.00	11775.00	11775.00
		(30.79)	(28.62)	(26.70)	(24.68)	(23.23)
e)	Interest on past capital	2159.38	4549.85	7121.46	9877.59	12859.85
		(5.65)	(11.06)	(16.15)	(20.70)	(25.37)
	Total fixed cost	16329.26	18719.73	21291.33	24047.47	27029.73
		(42.69)	(45.50)	(48.28)	(50.40)	(53.33)
8	Total cost	38247.54	41145.70	44098.16	47716.19	50688.59
		(100.00)	(100.00)	(100.00)	(100.00)	(100.00)

Figure in parentheses are percentage to total cost

Material cost contributed 4.91 per cent to 6.46 per cent of the total cost in the fifth year to first year respectively (Table.3). Total labour cost varied from 44.13 per cent to 46.70 per cent in the first to fifth year. Per cent share of interest on working capital was 1.75 per cent and risk margin and managerial cost contributed equally to the total cost was i.e 0.88 per cent at an overall level. While interest on past establishment cost was found to have positive relationship with the age during non-bearing years. Renal value of land was 30.79 per cent, 28.62 per cent, 26.70

per cent, 24.68 per cent and 23.23 per cent from first to fifth year, respectively. Rental value of land in the total cost declined with the increase age of plants. Total cost was estimated to Rs. 38247.54, Rs. 41145.70, Rs. 44098.16, Rs. 47716.19 and Rs. 50688.59 per hundred plants in the first year to fifth year respectively (Table 3).

In case of medium category, total cost was Rs.37643.71, Rs. 40884.44, Rs. 44108.56, Rs. 47315.67 and Rs. 50668.37 per hundred plants from the first to fifth year respectively (Table 4). The per cent share of FYM in total cost was maximum followed by plant protection chemicals and Bordeaux mixture. The total material cost varied from 5.21 per cent to 6.59 per cent from fifth to first year .Total labour cost in medium farm varied from 34.81 per cent in the fifth year to 41.69 per cent in first year. As expected interest on working capital exhibited positive relation with age while interest on fixed capital decreased with age of plantation.

In case of overall farms, it was Rs. 38391.13 Rs. 40998.07, Rs. 44050.47, Rs. 47271.17 and Rs. 50540.90 per hundred plants in the first to fifth year respectively (Table 5). Total variable cost in case of overall farms contributed 46.73 per cent in the first year and 57.67 per cent of total costs in the fifth year. Total fixed cost varied from 42.33 per cent to 53.27 per cent of the total costs in the first year to fifth year .The material cost was increased with the age of plants. The interest on working capital showed the positive relation with age while interest on fixed capital decreased with age of plantation.

Table 4. Maintenance cost during non- bearing stage of mango orchard in case of medium farms (Rupees/100 Plants)

Sr.No.	Item of cost	Age of plantation in years				
		1st year	2nd year	3rd year	4th year	5th year
1	**Material Cost**					
a)	Gap filling	204.92	-	-	-	-
		(0.54)				
b)	Bordeaux mixture	245.90	273.11	287.23	366.11	380.00
		(0.65)	(0.67)	(0.65)	(0.77)	(0.75)
c)	FYM	1557.38	1621.85	1693.62	1715.48	1720.00
		(4.14)	(3.97)	(3.84)	(3.63)	(3.39)
d)	Plant protection chemical	471.52	495.80	510.64	535.56	540.00
		(1.25)	(1.21)	(1.16)	(1.13)	(1.07)
	Total material cost	2479.71	2390.76	2491.49	2617.15	2640.00
		(6.59)	(5.85)	(5.65)	(5.53)	(5.21)
2	**Human labour**					
a)	Family labour	10342.21	10823.53	11125.53	11232.22	11270.00
		(27.47)	(26.47)	(25.22)	(23.74)	(22.24)
b)	Hired labour	5350.41	5764.71	5957.45	6106.69	6370.00
		(14.21)	(14.10)	(13.51)	(12.91)	(12.57)

	Total human labour	15692.62	16588.24	17082.98	17338.91	17640.00
		(41.69)	(40.57)	(38.73)	(36.65)	(34.81)
3	Interest on working capital	391.51	407.77	422.45	436.19	450.50
		(1.04)	(1.00)	(0.96)	(0.92)	(0.89)
4	Risk Margin	783.01	815.55	844.89	872.38	901.00
		(2.08)	(1.99)	(1.92)	(1.84)	(1.78)
5	Managerial cost	783.01	815.55	844.89	872.38	901.00
		(2.08)	(1.99)	(1.92)	(1.84)	(1.78)
6	Total Variable Cost	20129.87	21017.86	21686.70	22137.03	22532.50
		(53.47)	(51.41)	(49.17)	(46.79)	(44.47)
7	**Fixed cost**					
a)	Land revenue	31.00	31.00	31.00	31.00	31.00
		(0.08)	(0.08)	(0.07)	(0.07)	(0.06)
b)	Deprecation	2399.15	2399.15	2399.15	2399.15	2399.15
		(6.37)	(5.87)	(5.44)	(5.07)	(4.74)
c)	Interest on Fixed capital	944.93	944.93	944.93	944.93	944.93
		(2.51)	(2.31)	(2.14)	(2.00)	(1.86)
d)	Rental value of land	11775.00	11775.00	11775.00	11775.00	11775.00
		(31.28)	(28.80)	(26.70)	(24.89)	(23.24)
e)	Interest on past capital	2363.77	4716.51	7271.78	10028.57	12985.80
		(6.28)	(11.54)	(16.49)	(21.20)	(25.63)
	Total fixed cost	17513.85	19866.58	22421.86	25178.64	28135.87
		(46.53)	(48.59)	(50.83)	(53.21)	(55.53)
8	Total cost	37643.71	40884.44	44108.56	47315.67	50668.37
		(100.00)	(100.00)	(100.00)	(100.00)	(100.00)

Figure in parentheses are percentage to total cost

Table 5. Maintenance cost during non- bearing stage of mango orchard in case of overall farms (Rupees/100 Plants)

Sr.No.	Particulars	Age of plantation in years				
		1st year	2nd year	3rd year	4th year	5th year
1	**Material Cost**					
a)	Gap filling	156.55	-	-	-	-
		(0.41)				
b)	Bordeaux mixture	189.68	234.57	255.70	293.31	437.74
		(0.49)	(0.57)	(0.58)	(0.62)	(0.87)
c)	FYM	1483.09	1525.74	1607.60	1660.42	1645.47
		(3.86)	(3.72)	(3.65)	(3.51)	(3.26)
d)	Plant protection chemical	324.12	362.49	374.50	387.52	396.16
		(0.84)	(0.88)	(0.85)	(0.82)	(0.78)
	Total material cost	2153.43	2122.79	2237.81	2341.25	2479.37
		(5.61)	(5.18)	(5.08)	(4.95)	(4.91)

2	**Human labour**					
a)	Family labour	17255.96	17326.93	17595.35	17727.69	17698.39
		(44.95)	(42.26)	(39.94)	(37.50)	(35.02)
b)	Hired labour	1752.55	1892.40	1954.66	2119.40	2256.94
		(4.56)	(4.62)	(4.44)	(4.48)	(4.47)
	Total human labour	19008.50	19219.34	19550.01	19847.09	19955.33
		(49.51)	(46.88)	(44.38)	(41.99)	(39.48)
3	Interest on working capital	195.30	200.76	209.62	223.03	236.82
		(0.51)	(0.49)	(0.48)	(0.47)	(0.47)
4	Risk Margin	390.60	401.52	419.25	446.06	473.63
		(1.02)	(0.98)	(0.95)	(0.94)	(0.94)
5	Managerial cost	390.60	401.52	419.25	446.06	473.63
		(1.02)	(0.98)	(0.95)	(0.94)	(0.94)
6	Total Variable Cost	22138.43	22345.93	22835.94	23303.50	23618.78
		(57.67)	(54.50)	(51.84)	(49.30)	(46.73)
7	**Fixed cost**					
a)	Land revenue	31.00	31.00	31.00	31.00	31.00
		(0.08)	(0.08)	(0.07)	(0.07)	(0.06)
b)	Deprecation	1662.01	1662.01	1662.01	1662.01	1662.01
		(4.33)	(4.05)	(3.77)	(3.52)	(3.29)
c)	Interest on Fixed capital	636.79	636.79	636.79	636.79	636.79
		(1.66)	(1.55)	(1.45)	(1.35)	(1.26)
d)	Rental value of land	11775.00	11775.00	11775.00	11775.00	11775.00
		(30.67)	(28.72)	(26.73)	(24.91)	(23.30)
e)	Interest on past capital	2147.91	4547.35	7109.73	9862.89	12817.33
		(5.59)	(11.09)	(16.14)	(20.86)	(25.36)
	Total fixed cost	16252.70	18652.14	21214.52	23967.68	26922.13
		(42.33)	(45.50)	(48.16)	(50.70)	(53.27)
8	Total cost	38391.13	40998.07	44050.47	47271.17	50540.90
		(100.00)	(100.00)	(100.00)	(100.00)	(100.00)

Figure in parentheses are percentage to total cost

The depreciation cost and rental value of land was declined with the increase age of plants. While interest on past establishment cost was found to have positive relationship with the age during non-bearing years. Interest on past establishment cost was increased with the increase in the age of plants which was the main component responsible for highest per cent share of fixed cost in the total cost.

1.3 Cost During Bearing Stage of Mango Plantation in Different Farm Categories

In this section, an attempt has been made to find out maintenance cost of mango during bearing stage for marginal, small, medium and overall farm category (Table 6 to Table 9).

Table 6 revealed that in marginal farm category, total cost increase in the early and main bearing stage and was calculated to Rs. 56373.26 and Rs. 58275.84 per hundred plants and then gradually decreased to Rs. 54354.65 in the late bearing stage (Table 6). The fertilizers were maximum in main bearing stage Rs.2639.86 followed by late bearing stage Rs.2600.00 and early bearing stage Rs. 2419.35. The per cent share of FYM to total cost was maximum in main bearing stage (8.76%) followed by early bearing (8.56%) and late bearing stage (8.18%) respectively. Total labour cost increased from 36.46 per cent to 36.92 per cent from early to main bearing stage and thereafter it decreased to34.34 per cent. The share of variable cost showed an increasing trend up to the main bearing stage and thereafter it declined in the late bearing stage. However, fixed cost in percentage term to total cost exhibited declining trend up to the main bearing stage and then started increasing.

In small farmers, total cost varied from Rs. 55805.36 to Rs. 57532.91 per hundred plants from early to main bearing stage and then gradually deceased in Rs. 55832.37 per hundred plants respectively (Table 7). The per cent share of total material cost in total cost was maximum in main bearing stage (9.34%) followed by late bearing stage (9.26%) and early bearing stage (8.13%) in small farm category. Human labour varied from 39.80 per cent to 41.17 per cent from late bearing to early bearing stage. The share of variable cost in the total maintenance cost among different farm categories showed an increasing trend up to the main bearing stage and thereafter it started decreasing.

Table 6. Maintenance cost during bearing stage of mango orchard in case of marginal farms

(Rupees/100 Plants)

Sr.No.	Particulars	Bearing stages of mango		
		Early bearing	Main bearing	Late bearing
1	**Material Cost**			
a)	Bordeaux mixture	266.58	306.60	261.11
		(0.47)	(0.53)	(0.48)
b)	FYM	4935.48	5000.00	4444.44
		(8.58)	(8.76)	(8.18)
c)	Fertilizer	2419.35	2639.86	2600.00
		(4.29)	(4.53)	(4.78)
d)	Plant protection chemical	641.94	667.36	635.56
		(1.14)	(1.15)	(1.17)
	Total material cost	8263.35	8613.82	7941.11
		(14.66)	(14.78)	(14.61)
2	**Human labour**			
a)	Family labour	16572.94	15511.81	13222.22
		(29.40)	(26.62)	(24.33)
b)	Hired labour	3982.97	6003.47	5444.44
		(7.07)	(10.30)	(10.02)
	Total human labour	20555.91	21515.28	18666.67

Sr.No.	Particulars	Bearing stages of mango		
		Early bearing	Main bearing	Late bearing
		(36.46)	(36.92)	(34.34)
3	Interest on working capital	612.32	730.86	669.28
		(1.09)	(1.25)	(1.23)
4	Risk Margin	1224.63	1461.73	1338.56
		(2.17)	(2.51)	(2.46)
5	Managerial cost	1224.63	1461.73	1338.56
		(2.17)	(2.51)	(2.46)
6	Total Variable Cost	31880.85	33783.42	29887.24
		(56.55)	(57.97)	(54.99)
7	**Fixed cost**			
a)	Land revenue	31.00	31.00	31.00
		(0.05)	(0.05)	(0.06)
b)	Deprecation	1180.31	1180.31	1180.31
		(2.09)	(2.03)	(2.17)
c)	Interest on Fixed capital	446.68	446.68	446.68
		(0.79)	(0.77)	(0.82)
d)	Rental value of land	11775.00	11775.00	11775.00
		(20.89)	(20.21)	(21.66)
e)	Pro-rated establishment cost	11059.42	11059.42	11059.42
		(19.62)	(18.98)	(20.35)
	Total fixed cost	24492.42	24492.42	24467.42
		(43.45)	(42.03)	(45.01)
8	Total cost	56373.26	58275.84	54354.65
		(100.00)	(100.00)	(100.00)

Figure in parentheses are percentage to total cost

Table 7. Maintenance cost during bearing stage of mango orchard in case of small farms

(Rupees/100 Plants)

Sr.No.	Particulars	Bearing stages of mango		
		Early bearing	Main bearing	Late bearing
1	Material Cost			
a)	Bordeaux mixture	218.25	291.21	267.86
		(0.39)	(0.51)	(0.48)
b)	FYM	1704.76	2032.97	2023.81
		(3.05)	(3.53)	(3.62)
c)	Fertilizer	2092.38	2454.95	2345.24
		(3.75)	(4.27)	(4.20)
d)	Plant protection chemical	523.81	591.87	535.71
		(0.94)	(1.03)	(0.96)
	Total material cost	4539.21	5370.99	5172.62
		(8.13)	(9.34)	(9.26)

Sr.No.	Particulars	Bearing stages of mango		
		Early bearing	Main bearing	Late bearing
2	Human labour			
a)	Family labour cost	15422.22	13602.56	14444.44
		(27.64)	(23.64)	(25.87)
b)	Hired labour cost	7555.56	9561.54	7777.78
		(13.54)	(16.62)	(13.93)
	Total human labour	22977.78	23164.10	22222.22
		(41.17)	(40.26)	(39.80)
3	Interest on working capital	604.74	746.63	647.52
		(1.08)	(1.30)	(1.16)
4	Risk Margin	1209.48	1493.25	1295.04
		(2.17)	(2.60)	(2.32)
5	Managerial cost	1209.48	1493.25	1295.04
		(2.17)	(2.60)	(2.32)
6	Total Variable Cost	30540.67	32268.22	30567.69
		(54.73)	(56.09)	(54.75)
7	Fixed cost			
a)	Land revenue	31.00	31.00	31.00
		(0.06)	(0.05)	(0.06)
b)	Deprecation	1719.48	1719.48	1719.48
		(3.08)	(2.99)	(3.08)
c)	Interest on Fixed capital	644.40	644.40	644.40
		(1.15)	(1.12)	(1.15)
e)	Rental value of land	11775.00	11775.00	11775.00
		(21.10)	(20.47)	(21.09)
f)	Pro-rated establishment cost	11094.81	11094.81	11094.81
		(19.88)	(19.28)	(19.87)
	Total fixed cost	25264.69	25264.69	25264.69
		(45.27)	(43.91)	(45.25)
8	Total cost	55805.36	57532.91	55832.37
		(100.00)	(100.00)	(100.00)

Figure in parentheses are percentage to total cost

However, fixed cost in percentage term to total cost exhibited declining trend up to the main bearing stage and then started increasing.

Maintenance costs during bearing stage of mango in medium farm category were presented in Table 4.23. Expenditure incurred on FYM and fertilizers increased from 3.91 per cent to 4.31 per cent and 4.18 per cent to 5.17 per cent in early and main bearing stage and gradually decreased in late bearing stage 4.25 per cent and 5.07 per cent respectively. Total material costs varied from 9.89 per cent to 11.45 per cent of total cost of mango production. Interest on working capital increased in the early and main bearing stage and was calculated to Rs.756.76 and Rs. 917.94

and then decreased to Rs.Rs.843.50 in late bearing stage. The proportionate share of variable costs in the total maintenance cost showed increasing trends up to main bearing stage and thereafter started decreasing. Among the fixed costs, rental value of land and prorated establishment cost accounted for maximum proportion of the fixed costs. The proportion of fixed costs exhibited almost declining trend with the age. Table 8 revealed that total cost varied from Rs. 55565.96 to Rs. 57547.18 respectively.

At overall level, total cost varied from Rs. 56873.29 to Rs. 57861.69 from early to main bearing stage (Table 8). The costs were decreasing in the late bearing stages in all the categories of farms. In the initial years of development stage of the plants, the labour and material inputs were increasing with the growth of the plants up to maturity in the main bearing stage and after that productivity of mango start decreasing. Expenditure incurred on fertilizers and plant protection chemicals ranged from 3.75 to 4.62 per cent and 1.01 to 1.17 per cent respectively and the per cent share of total material cost in total cost ranged between 11.72 to 12.27 per cent exhibited positive relationships to age at overall level. Labour use almost doubled during the peak period of productivity phase i.e. in the main bearing stage as compared to early bearing stage and decreased in the late bearing stage. Table 8 revealed the per cent share of labour in total cost varied between 35.75 to 38.10 per cent in the various stages of mango. Risk margin and managerial costs varied from 2.35 per cent to 2.71 per cent at overall level and indicated positive relationship with age of plantation. Among fixed costs, rental value of land and prorated establishment cost accounted for maximum proportion of the fixed costs in all the categories of farms.

Table 8. Maintenance cost during bearing stage of mango orchard in case of medium farms

(Rupees/100 plants)

Sr.No.	Particulars	Bearing stages of mango		
		Early bearing	Main bearing	Late bearing
1	**Material Cost**			
a)	Bordeaux mixture	317.28	333.96	295.61
		(0.57)	(0.58)	(0.55)
b)	FYM	2171.22	2481.20	2263.51
		(3.91)	(4.31)	(4.25)
c)	Fertilizer	2324.48	2974.74	2702.70
		(4.18)	(5.17)	(5.07)
d)	Insecticide	680.83	796.99	729.73
		(1.23)	(1.38)	(1.37)
	Total material cost	5493.81	6586.89	5991.55
		(9.89)	(11.45)	(11.24)
2	Human labour			
a)	Family labour	10440.76	8417.54	6119.09
		(18.79)	(14.63)	(11.48)

b)	Hired labour	9661.46	11771.93	10878.38
		(17.39)	(20.46)	(20.41)
	Total human labour	20102.21	20189.47	16997.47
		(36.18)	(35.08)	(31.89)
3	Interest on working capital	757.76	917.94	843.50
		(1.36)	(1.60)	(1.58)
4	Risk Margin	1515.53	1835.88	1686.99
		(2.73)	(3.19)	(3.16)
5	Managerial cost	1515.53	1835.88	1686.99
		(2.73)	(3.19)	(3.16)
6	Total Variable Cost	29384.84	31366.07	27122.15
		(52.88)	(54.50)	(50.88)
7	**Fixed cost**			
a)	Land revenue	31.00	31.00	31.00
		(0.06	(0.05	(0.06)
b)	Deprecation	2399.15	2399.15	2399.15
		(4.32)	(4.17)	(4.50)
c)	Interest on Fixed capital	944.93	944.93	944.93
		(1.70)	(1.64)	(1.77)
d)	Rental value of land	11775.00	11775.00	11775.00
		(21.19)	(20.46)	(22.09)
e)	Pro-rated establishment cost	11031.04	11031.04	11031.04
		(19.85)	(19.17)	(20.69)
	Total fixed cost	26181.11	26181.11	26181.11
		(47.12)	(45.50)	(49.12)
8	Total cost	55565.96	57547.18	53303.26
		(100.00)	(100.00)	(100.00)

Figure in parentheses are percentage to total cost

Table 9. Maintenance cost during bearing stage of mango orchard in case of overall farms

(Rupees/100Plants)

Sr.No.	Particulars	Bearing stages of mango		
		Early bearing	Main bearing	Late bearing
1	Material Cost			
i)	Bordeaux mixture	271.20	309.16	280.35
		(0.48)	(0.53)	(0.51)
ii)	FYM	3721.42	3448.71	3284.65
		(6.50)	(5.95)	(5.88)
iii)	Fertilizer	2131.35	2672.29	2696.36
		(3.75)	(4.62)	(4.86)
iv)	Plant protection chemical	574.02	678.74	650.19
		(1.01)	(1.17)	(1.17)
	Total material cost	6697.99	7108.90	6911.55
		(11.74)	(12.27)	(12.41)

Sr.No.	Particulars	Bearing stages of mango		
		Early bearing	Main bearing	Late bearing
2	**Human labour**			
a)	Family labour	14995.79	13076.79	12137.51
		(26.30)	(22.58)	(21.71)
b)	Hired labour	6670.10	8585.11	7740.99
		(11.80)	(14.86)	(14.04)
	Total human labour	21665.90	21661.90	19878.50
		(38.10)	(37.44)	(35.75)
3	Interest on working capital	668.40	784.70	732.63
		(1.18)	(1.36)	(1.32)
4	Risk Margin	1336.81	1569.40	1465.25
		(2.35)	(2.71)	(2.65)
5	Managerial cost	1336.81	1569.40	1465.25
		(2.35)	(2.71)	(2.65)
6	Total Variable Cost	31705.91	32694.31	30379.92
		(55.72)	(56.50)	(54.64)
7	**Fixed cost**	0.00	0.00	0.00
a)	Land revenue	31.00	31.00	31.00
		(0.05)	(0.05)	(0.06)
b)	Deprecation	1662.01	1662.01	1662.01
		(2.94	(2.88	(3.02
c)	Interest on Fixed capital	636.79	636.79	636.79
		(1.13	(1.10	(1.16
d)	Rental value of land	11775.00	11775.00	11775.00
		(20.71	(20.35	(21.22
e)	Pro-rated establishment cost	11062.59	11062.59	11062.59
		(19.46	(19.12	(19.93
	Total fixed cost	25167.38	25167.38	25156.44
		(44.28	(43.50	(45.36
8	Total cost	56873.29	57861.69	55536.36
		(100.00	(100.00	(100.00

Figure in parentheses are percentage to total cost

1.4 Returns from Mango cultivation

Table 10 depicts that the average production on marginal farm was 51.43 quintals during early bearing stage that increased to 55.70 during main bearing stage and then decreased to 51.88 quintal in late bearing stage and in case of small and medium farm, average production increased from 50.50 quintals to 54.07 quintals and 49.71 quintal to 53.33 quintal in early to main bearing stage and then gradually decreased to 50.79 quintals and 49.84 quintals in late bearing stage, respevtively. It was observed that the average production was recorded highest in case of marginal farms i.e. 53.00 quintals followed by small farmers (51.79 quintals) and medium farmers (50.96 quintals). At overall level, the average production was 52.10 quintals.

Table 10. Farm category wise average production from different age group of mango plants in the sampled households

(Quintals)

Farm category	Average production of mango /100 plants			
	Early bearing	Main bearing	Late bearing	Average production
Marginal	51.43	55.70	51.88	53.00
Small	50.50	54.07	50.79	51.79
Medium	49.71	53.33	49.84	50.96
Overall	50.70	54.59	51.02	52.10

Table 11. Farm category wise average gross returns from different age groups of mango in the sampled households

(Rupees)

Farm category	Average gross return of mango /100 plants			
	Early bearing	Main bearing	Late bearing	Average gross return
Marginal	180000.00	194957.98	181564.25	185507.41
Small	176750.00	189235.27	177777.78	181254.35
Medium	173987.27	186666.67	174442.68	178365.54
Overall	177446.66	191064.70	178558.89	182356.75

The average gross returns were highest in case of marginal farms i.e. Rs. 185507.41 followed by small farms Rs. 181254.35 and medium farms Rs. 178365.54 (Table 11). At overall level, gross returns was Rs 177446.66 during early bearing stage that increased to Rs. 191064.70 in main bearing stage and then decreased to Rs. 178558.89 in late bearing stage. Average gross returns of mango at overall level were Rs. 182356.75

Conclusions

The study of mango farming system highlights the socio-economic aspects objectives, priorities, cultivation practices and problems encountered by the orchardists. Maintenance cost of bearing mango per hundred plants at overall level showed an increase in cost up to main bearing stage and thereafter, it started decreasing which is due to the decrease in production and lead to lesser labour requirement. The proportion of variable cost ranged from 55.72 per cent to 56.50 per cent upto main bearing stage. The per cent share of fixed cost varied from 43.50 per cent to 45.36 per cent in the total cost in different ages of plantation. Cost incurred on material inputs ranged from 11.74 per cent to 12.41 per cent while human labour cost was around 35.75 per cent to 38.10 per cent of the total cost for the different age groups. Average production varied from 50.96 quintals to 53.00 quintals per 100 plants and net returns varied from Rs. 178365.54 to Rs. 185507.41 in different farm categories. The orchardists should be trained properly in performing various operations in the orchard, especially timely application of the fertilizers & insecticides and other improved management practices. The inputs like fertilizers, plant protection chemicals should be made available to the orchardists in time and on subsidized rates.

References

Acharya SS and Agarwal NL. 1987. Agricultural Marketing in India. pp. 309-24.

Akcay Y and Uzunoz M. 2005. An investment analysis of peach and cherry growing in Turkey. Journal of Applied Science 5:1665–68.

Chand R. 1994. Economics of perennial crops: Some methodological issues. Indian Journal of Agricultural Economics 49:1420–30.

Directorate of Economics and Statistics. 2020. Statistical Outlines of Himachal Pradesh 2020. Directorate of Economics and Statistics, Government of Himachal Pradesh.

National Horticulture Board. 2016. Horticultural Data Base 2016. National Horticulture Board, New Delhi.

8

Conservation Agriculture and Soil Health

Abhay Kumar Sinha, Vinod Gupta, Neerja Sharma and Saurav Gupta

Krishi Vigyan Kendra Samba, S.K. University of Agricultural Sciences & Technology of Jammu, Jammu and Kashmir, India

Conservation agriculture (CA) has **potential to changes soil physical, chemical and biological soil quality parameters** compared to conventional tillage (CT) systems. CA can also affect the functional diversity of soil microbes that essential for improved soil quality, crop production and many ecosystem services.

Conservation Agriculture is a farming system that promotes minimum soil disturbance (i.e. no tillage), maintenance of a permanent soil cover, and diversification of plant species. It enhances biodiversity and natural biological processes above and below the ground surface, which contribute to increased water and nutrient use efficiency and to improved and sustained crop production.

Three principles of Conservation Agriculture

Minimum mechanical soil disturbance

(i.e. no tillage) through direct seed and/or fertilizer placement.

Permanent soil organic cover

(at least 30 percent) with crop residues and/or cover crops.

Species diversification

through varied crop sequences and associations involving at least three different

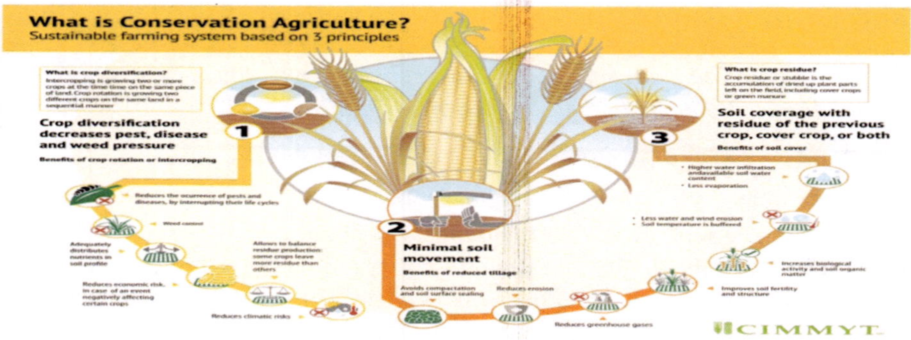

How did conservation agriculture originate?

The term "conservation agriculture" was coined in the 1990s, but the idea to minimize soil disturbance has its origins in the 1930s, during the Dust Bowl in the United States of America.

CIMMYT pioneered no-till training programs and trials in the 1970s, in maize and wheat systems in Latin America. In the 1980s this technique was also used in agronomy projects in South Asia.

CIMMYT began work with conservation agriculture in Latin America and South Asia in the 1990s and in Africa in the early 2000s. Today, these efforts have been scaled up and conservation agriculture principles have been incorporated into projects such as CSISA, FACASI, MasAgro, SIMLESA, and SRFSI.

Farmers worldwide are increasingly adopting conservation agriculture. In the 2015/16 season, conservation agriculture was practiced on about 180 mega hectares of cropland globally, about 12.5% of the total global cropland — 69% more than in the 2008/2009 season.

Is conservation agriculture organic?

Conservation agriculture and organic farming both maintain a balance between agriculture and resources, use crop rotation, and protect the soil's organic matter.

However, the main difference between these two types of farming is that organic farmers use a plow or soil tillage, while farmers who practice conservation agriculture use natural principles and do not till the soil. Organic farmers apply tillage to remove weeds without using inorganic fertilizers.

Conservation agriculture farmers, on the other hand, use a permanent soil cover and plant seeds through this layer. They may initially use inorganic fertilizers to manage weeds, especially in soils with low fertility. Over time, the use of agrichemicals may be reduced or slowly phased out.

How does conservation agriculture differ from climate-smart agriculture?

While conservation agriculture and climate-smart agriculture are similar, their purposes are different. Conservation agriculture aims to sustainably intensify smallholder farming systems and have a positive effect on the environment using natural processes. It helps farmers to adapt to and increase profits in spite of climate risks.

Climate-smart agriculture aims to adapt to and mitigate the effects of climate change by sequestering soil carbon and reducing greenhouse gas emissions, and finally increase productivity and profitability of farming systems to ensure farmers' livelihoods and food security in a changing climate. Conservation agriculture systems can be considered climate-smart as they deliver on the objectives of climate-smart agriculture.

Conservation agriculture and its components have been associated with many benefits including greater soil water storage, decreased erosion and in some instances, greater yield and net farm income. These benefits have led to the identification of CA as an important tool to help ensure future food production and help buffer agricultural productivity against extreme climate events, such as drought and heat waves, which are likely to increase in frequency under climate change (FAO, 2019). One of the key drivers of the improvements observed under CA is the greater soil organic matter (SOM), particularly at the surface of the profile, and the associated improvements in soil structural stability, fertility, and biological diversity relative to conventional agricultural systems.

The comprehensive summary of our knowledge regarding the effect of CA and its components (i.e., NT, residue retention and crop diversification) on SOM by conducting a narrative review of the peer reviewed research published worldwide over the past 30 years. The situations under which gains, no change or even losses of SOM have been observed when using CA systems will be examined and discussed. The subsequent changes in soil physical, chemical and biological properties stemming from changes in to SOM will also be explored and the ultimate impact of this on yield and yield stability in different environments and soil types examined. Some of the agronomic, social and economic challenges associated with the adoption of CA will be explored and potential opportunities and strategies required increasing the adoption of CA discussed.

Effect of CA on SOC (soil organic carbon)

The amount of organic matter present in a soil is the net difference between organic matter inputs (biomass) and losses (erosion, decomposition, leaching). The extent and direction that CA affects SOM (typically measured via SOC) is thus a function of how it impacts these inputs and losses. In --comparison to conventional agricultural systems, CA makes modifications to tillage practices, residue management and crop/nutrient management. Each of these modifications can influence SOC, as discussed below.

Tillage Management

It has been well-established that tilling the soil leads to losses of SOC as cultivation breaks up the soil and exposes organic matter previously protected within soil macro aggregates to microbial decay. Cultivation also incorporates and fragments plant material, increasing its vulnerability to microbial attack. Decreasing the amount of tillage or introducing no-till (NT) thus has the potential to decrease the amount of SOC lost from the profile by decreasing the turnover rate of macro aggregates, increasing the physical protection of particulate organic material, and reducing soil to residue contact. For example, in a Brazilian Acrisol under cereal cropping, turnover times of 17 vs. 36 years were observed in conventional and CA, respectively, due to the reduced disturbance in the NT CA system.

However, it should be noted that in some instances tillage has also been observed to increase SOC stores relative to CA. In areas where full inversion tillage is carried out, residues may be buried in a region where poor soil aeration can limit decomposition (relative to the soil surface), particularly under cool, moist climatic conditions.

Residue Management

Crop residues can be defined as biomass remaining on the soil's surface after harvest. In some systems, linear increases in SOC stocks can be observed with increasing rates of residue addition. Thus, in CA systems, which emphasize the retention of residues on the soil surface, greater residue input can potentially lead to greater SOC storage. However, where residue production is low e.g., due to low soil fertility or the presence of soil constraints, there can be insufficient residue retention under CA to positively impact SOC. Indeed, in some situations CA can even lead to a reduction in plant growth (e.g., due to lower soil temperatures, compaction or greater disease/weed/pest pressure) and this can also lead to decreases in residue inputs and lower or no SOC sequestration. In areas with low fertility, integrated nutrient management is essential to ensure a buildup of SOC and the success of CA systems. Low fertility results in low crop yields, which leads to reduced organic matter input and hence, lower SOC. This cycle can only be broken by the addition of nutrients to the soil system via external fertilizer input (organic or inorganic) and/or the incorporation of legumes into cropping rotations.

Crop Rotation

Different crops may have different effects on the quantity, quality, and periodicity of C inputs and can modify the soil in different ways (e.g., rates of water extraction, nutrient use), which can influence mineralization rates and the growth of subsequent crops . Thus, differences in crop rotation between CA and conventional agricultural systems also have the potential to impact SOC values. The elimination of monocultures and incorporation of plant species into rotations that return greater amounts of residue to the soil are often associated with greater SOC stock in CA systems. In some systems, root input, in particular, has been found to be important. Where legumes are included, these can also add additional N to the system, which can enhance soil fertility and subsequent crop biomass production. The maintenance of residue cover can also decrease processes that can limit the growth and biomass production of the main crop, such as erosion, nutrient leaching, and weeds, pests and.

Net Effects of CA on SOC

The exact magnitude of the difference in SOC between CA and conventional agricultural systems varies greatly and is influenced by many factors, including climate, soil type; baseline SOC, crop management, time since management change, and sampling depth and methodology. This means that estimates of the magnitude of the change following conversion to CA vary widely. For example, estimates ranging from −0.15 Mg/ha/year in areas such as the Midwestern USA to +0.93 Mg/ha/years in tropical Brazil have been recorded. However, it is clear that in regions where soil and climatic conditions are favorable for biomass production and where negative yield impacts are not observed, then CA systems will often have higher amounts of SOC relative to conventionally managed systems, particularly in the surface of the soil profile. Although, the large range in the sequestration rates observed indicates that the magnitude of change is likely to be highly site specific. It should also be noted that most of the literature regarding the effect of CA on soil properties has focused on NT and residue retention v conventional tillage and residue removal. Fewer studies have also considered the impact of species diversification and the inclusion of practices such as cover crops and intercropping on SOC change. Where these are carried out they are often found to lead to greater increases in SOC than NT and stubble retention alone, and the effect of management changes on soil properties would thus often be greater when all elements of the CA system, which may have interactive and synergic effects, are incorporated together.

Advantages of Conservation Agriculture

Conservation Agriculture is 20 to 50 percent less labor intensive and thus contributes **to reducing greenhouse gas emissions through lower energy inputs and improved nutrient use efficiency**. At the same time, it stabilizes and protects soil from breaking down and releasing carbon to the atmosphere.

Conservation agriculture is generally a "win-win" situation for both farmers and the environment. Yet many people intimately involved with worldwide food production have been slow to recognize its many advantages and consider it to be a viable alternative to conventional agricultural practices that are having obvious negative impact on the environment. Much of this has to do with the fact that conservation agriculture requires a new way of thinking about agricultural production in order to understand how one could possibly attain higher yields with less labour, less water and fewer chemical inputs. In spite of these challenges, conservation agriculture is spreading to farmers throughout the world as its benefits become more widely recognized by farmers, researchers, scientists and extensionists alike.

Specifically, conservation agriculture (CA) increases the productivity of:

Land - Conservation agriculture improves soil structure and protects the soil against erosion and nutrient losses by maintaining a permanent soil cover and minimizing soil disturbance. Furthermore, CA practices enhance soil organic matter (SOM) levels and nutrient availability by utilizing the previous crop residues or growing green manure/ cover crops (GMCC's) and keeping these residues as a surface mulch rather than burning. Thus, arable land under CA is more productive for much longer periods of time.

Labour - Because land under no-till is not cleared before planting and involves less weeding and pest problems following the establishment of permanent soil cover/ crop rotations, farmers in Ghana reported a 22% savings in labour associated with maize production. Similar reductions in labour requirements have been reported with no-till rice-wheat systems in South Asia and various CA technologies in South America. Much of the reduced labour comes from the absence of tillage operations under CA, which use up valuable labour days during the planting season.

Water - Conservation agriculture requires significantly less water use due to increased infiltration and enhanced water holding capacity from crop residues left on the soil surface. Mulches also protect the soil surface from extreme temperatures and greatly reduce surface evaporation, which is particularly important in tropical and sub-tropical climates. In Sub-Saharan Africa, as with other dry land regions, the benefits of conservation agriculture are most salient during drought years, when the risk of total crop failure is significantly reduced due to enhanced water use efficiency.

Nutrients - Soil nutrient supplies and cycling are enhanced by the biochemical decomposition of organic crop residues at the soil surface that are also vital for feeding the soil microbes. While much of the nitrogen needs of primary food crops can be achieved by planting nitrogen-fixing legume species, other plant essential nutrients often must be supplemented by additional chemical and/or organic fertilizer inputs. In general, soil fertility is built up over time under conservation agriculture, and fewer fertilizer amendments are required to achieve optimal yields over time.

Soil biota - Insect pests and other disease causing organisms are held in check by an abundant and diverse community of beneficial soil organisms, including predatory wasps, spiders, nematodes, springtails, mites and beneficial bacteria and fungi, among other species. Furthermore, the burrowing activity of earthworms and other fauna create tiny channels or pores in the soil that facilitate the exchange of water and gases and loosen the soil for enhanced root penetration.

Economic benefits - Farmers using CA technologies typically report higher yields (up to 45-48% higher) with fewer water, fertilizer and labour inputs, thereby resulting in higher overall farm profits. In Paraguay, net farm income of no-till (NT) farming on large-scale commercial farms increased from $2,3467 to $32,608 more than farms using conventional tillage over a 10 year period. The economic benefits of NT and other conservation agriculture technologies, more than any other factor, have lead to widespread adoption among both large- and small-scale farmers throughout the world.

Impact on Yield

Where improvements in soil properties are observed under a system of CA management, there are many instances where this has been associated with yield increases. Change in yield between conventional agricultural systems and those incorporating CA practices (NT + residue retention with or without crop diversification, as indicated by the cropping system).

The possibility for early planting is also greater in CA systems due to the reduced land preparation requirements, which may result in more efficient use of rainfall and reduced risk of crop failure due to terminal heat stress. Improvements in soil fertility and/or reductions in pests/diseases as a result of the inclusion of legumes and diversified species into cropping rotations have also been reported to contribute to yield increases in some instances.

Conclusion

It is clear that when CA systems are well-designed and adapted to local conditions, they can improve the SOC content of many soils compared to conventional agricultural systems and that this can lead to significant improvements in soil physical, chemical, and biological properties and productive capacity. However, to increase the adoption of CA worldwide, it is critical that the system be adapted to specific climates, soil types and communities, particularly considering the farmer's investment capacity, and the availability of resources.

References

Alam MK, Islam MM, Salahin N, Hasanuzzaman, M (2014) Effect of Tillage Practices on Soil Properties and Crop Productivity in Wheat Mungbean-Rice Cropping System under Subtropical Climatic Conditions. The Scientific World Journal 1-15.

Bhattacharyya R, Prakash V, Kundu S, Gupta HS (2006) Effect of tillage and crop rotations on pore size distribution and soil hydraulic conductivity in sandy clay loam soil of the Indian Himalayas. Soil Tillage Research 82: 129-140.

Bhushan L, Ladha JK, Gupta RK, Singh S, Tirol-Padre A, Saharawat YS, Gathala M, Pathak H (2007) Saving of water and labor in rice-wheat systems with no-tillage and direct seeding technologies. Agronomy Journal 99: 1288-1296

DangYP, Seymour NP, Walker SR, Bell MJ, Freebairn DM (2015) Strategic tillage in no-till farming systems in Australia's northern grains growing regions: I. Drivers and implementation. Soil Tillage Research 152:104–114.

Das TK, Bhattacharyya R, Sudhishri S, Sharma AR, Saharawat YS, Bandyopadhyay KK, Seema S, Bana RS, Aggarwal P, Sharma RK, Bhatia A, Singh G, Datta SP, Kar A, Singh B, Singh P, Pathak H, Vyas AK, Jat ML (2014) Conservation agriculture in an irrigated cotton–wheat system of the western Indo-Gangetic Plains: crop and water productivity and economic profitability. Field Crops Research 158: 24–33.

9

Control of Plant Diseases Under Organic Crop Production

Saurav Gupta,Vinod Gupta,Neerja Sharma and A.K. Sinha

KVK Samba, SKUAST-Jammu, Jammu and Kashmir, India

Plant diseases create challenging problems in commercialagriculture and pose real economic threats to bothconventional and organic farming systems. Plantpathogens are difficult to manage for several reasons.First of all, plant pathogens are hard to identify becausethey are so small. The positive identification of apathogen often requires specialized equipment andtraining, and in some cases accurate diagnosis in thefield is difficult.Plant pathogens are constantly changing and mutating,resulting in new strains and new challenges togrowers. Disease management is complicated by the presenceof multiple types of pathogens. For any crop thegrower must deal with a variety of fungi, bacteria, viruses,and nematodes. This situation is even more complicatedfor organic crop production because they usuallyproduce a wide array of crops and are prohibitedfrom applying conventional synthetic fungicides.The world market continues to be extremely competitiveand continues to require that growers supplyhigh-quality, disease-free produce with an acceptableshelf life. Disease management is therefore a critical considerationin organic crop production.In an organic system, it is appropriate to developdisease-control strategies that have an ecological basis.For example, insofar as possible the organic systemshould encourage the growth and diversity of soilinhabitingand epiphytic (plant surface dwelling)microorganisms that have the potential to exert beneficialand pathogen-antagonistic influences. An increasein the genetic diversity of the crop host rotation isanother management step that incorporates ecologicalconsiderations. The integration of disease managementdecisions with insect and weed control and general productionpractices is another step consistent with thisapproach.

Plant Disease Diagnostics

The first step in any management decision regarding disease control is to determine which diseases and pathogens are causing the problem. Accurate and timely diagnosis of plant diseases is an essential component of integrated disease control in organic and conventional systems. Disease diagnosis is enhanced when all professionals, including the grower, field personnel, pest control advisor, consultant, and extension personnel, work together to ascertain the cause of the problem. Often, field identification is impossible and samples must be submitted to a qualified laboratory for analysis. Once a diagnosis has been determined, growers and other decision makers can settle on appropriate steps to take to manage the problem.

Site Selection

Before planting crops, a grower should carefully plan out planting and crop rotation strategies to avoid insofar as possible any known problem areas. A grower can incur significant losses if he or she plants susceptible crops in a field known to be infested with persistent soilborne pathogens. Plant-pathogenic fungi such as *Armillaria, Fusarium*(the wilt-causing species), *Plasmodiophora, Sclerotium,* and *Verticillium*are true soil inhabitants and will persist in soil for many years, even in the absence of a plant host. Because not all fields are infested with these fungi, the grower is advised to select a planting site away from such fields. Soilborne fungi such as *Phytophthora, Pythium,* and *Rhizoctonia*often are much more widespread, so site selection might be less of an option in avoiding these organisms. There are also other planting situations that create risks that should be avoided. Pastures, foothills, riverbanks, grasslands, and other areas that support weeds and natural vegetation often are reservoirs of pathogens that cause virus and viruslike diseases. The vectors that carry such pathogens also can be found in these high-risk areas and often migrate into production fields. For example, the aster yellows phytoplasma and its leafhopper vector can be found in weedy grasslands in coastal California. Once the grassland vegetation dries up in the summer, the leafhoppers migrate into nearby lettuce or celery fields, resulting in aster yellows disease in these fields. Consider pertinent environmental factors when selecting a planting site. Crops planted very close to the seacoast tend to be more at risk from downy mildew diseases as a result of the increased and persistent humidity. Just a few miles inland from the ocean, however, humidity can be significantly lower, decreasing the disease pressure for downy mildew. An understanding of soil factors is critical in avoiding some root and crown diseases. A site that has well-drained, sandy soil reduces the risk of damping-off and root rot for sensitive crops such as spinach. For any site selection decision, careful and detailed record keeping is essential. As a grower, you should keep notes on previous soil borne disease problems associated with certain fields, the position of fields in relation to other key areas (weed reservoirs), the environmental characteristics of importance for each location, and the nature of soil, water, and other physical features of each site.

Exclusion

The practice of keeping out any materials or objects that are contaminated with pathogens or diseased plants and preventing them from entering the production system is known as *exclusion.* For some diseases, seed borne pathogens are a primary means of pathogen dissemination. Growers should purchase seed that has been tested and certified to be below a certain threshold infestation level or that has been treated to reduce pathogen infestation levels. Note that the designation "pathogen free seed" really is not a valid term because it is not possible to know whether a seed lot is, in its entirety, absolutely free of all pathogens. Seed tests only examine representative samples, but in most cases the tests are accurate enough to give a true picture of the risk of diseases initiated by seedborne pathogens. If a grower produces or purchases transplants, they too should be as free as possible of pathogen contamination (where the pathogen is present on the plant but has not yet caused visible symptoms) and from disease (where symptoms are actually visible). For greenhouse crop production or the production of transplants, all materials should be clean and free of pathogens. By using clean or new pots, trays, and soilless potting mix, a grower can prevent the introduction of soilborne pathogens into the greenhouse system. The recycling of potting mixes is strongly discouraged, and pots and trays should be reused only if they are properly cleaned with steam, bleach, or other disinfectants. Soil and water can harbor pathogens as well. Take care to see that no infested soil or water is introduced into uninfested areas. Tomato bushy stunt virus of lettuce, tomato, and other crops is found in river, flood, and runoff waters. Growers who have dredged up soil from ditches and dispersed it onto fields have found that their fields can become infested with the virus and subsequent plantings can be severely diseased. Water draining from fields can carry a number of pathogens, and growers should not recycle or reuse it without carefully considering potential risks and then taking appropriate safety precautions. Soil adhering to tractor equipment and implements can spread soilborne pathogens from infested fields into clean fields. It is a good idea to reduce the off-site movement of these infested materials as much as possible.

Cultural Practices

There are a number of cultural practices that a grower should consider when designing an integrated disease control system. As a general approach, growers should take steps to grow vigorous, high-quality plants using the best farming practices possible. Listed below are some specific cultural practices that can help to manage diseases. Crop rotation is an important consideration in disease management. Rotation using diverse crops, inclusion of cover crops, and appropriate use of fallow (hostfree) periods all can contribute to the reduction of inoculum levels for soilborne pathogens and the increase of diversity in soil microflora. In contrast, consecutive plantings of the same crop in the same field often lead to increases in soilborne pathogens. Too little crop rotation in a given area can also simulate a onoculture effect that might increase foliar diseases. Recent research has shown that certain plants, besides being revenue-generating crops, also have a suppressive effect on diseases. For example, after broccoli and other crucifer crops are harvested and the plant residue is plowed into the soil, the decomposition of the broccoli stems and leaves releases natural chemicals that can significantly reduce the number of *Verticilliumdahliae*microsclerotia. This broccoli effect can be an important consideration in crop rotation strategies. Some cover crops (mustards, sudangrass) might also share this beneficial effect and could be considered in the crop rotation scheme. It is important to remember that while rotations with non-susceptible plants and cover crops may help reduce soilborne pathogen numbers, significant decreases in such populations are likely to take many seasons. When devising a crop rotation strategy, a grower should also be aware of which crops and cover crops might increase disease problems. A vetch cover crop, if planted into a field with a history of lettuce drop, can greatly increase the number of infective sclerotia of *Sclerotinia minor.* Vetch is a known host of root-knot nematode (*Meloidogyne*species) and also might increase soil populations of *Pythium*and *Rhizoctonia*dampingoff fungi. While oilseed radish could be a potential trap crop for cyst nematode (*Heterodera*species), as a cover crop it is a host of root-knot nematode and the clubroot fungus (*Plasmodiophorabrassicae*). There are many factors to consider in regard to planting a crop. Timing can be an important question. If cauliflower is planted into *Verticillium*-infested fields in the spring or summer, it is likely to experience disease and possible crop loss. However, if cauliflower is planted into the same fields in the late fall or winter it will exhibit no Verticillium wilt symptoms, presumably because the soil temperatures are too cool to allow the fungus to develop and cause significant disease. Disease can also be influenced by steps taken prior to and during the planting process. Tillage procedures should reduce plant residues left from previous crops Proper preparation of the field and the subsequent raised beds should reduce problems in areas that are subject to poor drainage, pooling of water, and other conditions that favor pathogens. Soil and bed preparation should result in good soil tilth so that seed or transplants are placed in a soil that favors plant development. Planting depth for seed or transplants should be tailored to enhance seed emergence or transplant establishment. Poor soil preparation can result in stressed and exposed plants and increased damping-off problems due to soil fungi. Irrigation management is clearly an important factor when it comes to disease control. Regardless of the irrigation method a grower chooses (furrow, sprinkler, or drip), timing and duration of irrigations should satisfy crop water requirements without allowing for excess water. Overwatering greatly favours most soilborne pathogenic fungi. For most foliar diseases, overhead sprinkler irrigation enhances pathogen survival and dispersal and disease development.Bacterial foliar diseases are particularly dependent upon rain and sprinkler irrigation. A grower should consider limiting or eliminating sprinkler irrigation if foliar diseases are problematic for a specific crop or field. The selection and application of fertilizers, in a few documented situations, can significantly influence disease development. For example, the use of the nitrate form of nitrogenous fertilizers can increase the severity of lettuce corky root disease. The excessive use of nitrogen fertilizers can result in leaf growth that is overly succulent and more susceptible to some diseases. On the other hand, liming the soil to raise pH levels can reduce symptom expression for clubroot disease of crucifers. In general, however, fertilizer management is not directly related to disease control. *Field sanitation* is the removal or destruction of diseased plant residues. In some field situations, sanitation is an appropriate step for managing diseases. Once lettuce has been harvested, for example, the remaining plants can act as a reservoir for lettuce mosaic virus. Sanitation in this case would include plowing down the old plants. Lettuce drop, caused by the fungus *Sclerotinia minor*, occurs when sclerotia develop on lettuce plant residues and remain in the top few inches of soil. One form of sanitation involves deep plowing in which moldboard plows invert the soil and bury sclerotia. Note that this procedure is effective only if sclerotia are low to moderate in number. Sanitation measures are more commonly applied in greenhouse situations. The removal of dead or dying transplants can help reduce inoculum that could otherwise spread to adjacent transplants. The removal of senescent tomato or cucumber plants might reduce (though not prevent) the spread of *Botrytis* spores. *Roguing*is a special form of plant sanitation that involves the physical removal of diseased plants from the field. While not applicable in many situations, researchers have shown that for sclerotia-forming fungi (such as *Sclerotinia minor* on lettuce) the regular removal of diseased plants can gradually reduce the overall number of sclerotia in fields. The management of other pests is a cultural control that could greatly influence the development of plant diseases. In particular, virus disease management is more effective when weeds and insects are also controlled. Weeds are known reservoirs of a number of viral and bacterial pathogens.

Soil solarization is the use of plastic tarps placed on the soil surface to increase soil temperatures to a level that kills soil borne pathogens, weeds, and other crop pests. Soil solarization works best in areas with acceptably high summer temperatures. These temperatures generally do not occur in California's coastal regions. Soil solarization will not eradicate a pathogen from a field, but it may lower pathogen populations. Soil flooding is a related though seldom-used means of creating conditions—in this case, saturated soil over an extended period—that might result in a decline of soil borne pathogens. Finally, the ability to manipulate environmental conditions in a greenhouse vegetable transplant or production system can be used to help control diseases. Botrytis diseases can be better managed if warm, humid air is vented out of the greenhouse. Because rain is not a factor in greenhouses, many bacterial foliar diseases can be virtually eliminated if drip irrigation or sub-irrigation systems are used.

Incorporation of composts into soils is a fundamental cultural practice in organic production. Composts benefit the soil's fertility and condition in a number of ways, and also undoubtedly benefit disease management in some way. However, research studies and empirical data that clearly document any disease control benefits resulting from field-application of compost are lacking. Despite this lack of information on disease control, composts should be added to farmed soils in order to increase soil microflora diversity and populations.

Use of Resistant Cultivars

One of the most important components in an integrateddisease control program is the selection and planting ofcultivars that are resistant to pathogens. The term *resistance*usually describes the plant host's ability to suppressor retard the activity and progress of a pathogenicagent, which results in the absence or reduction ofsymptoms. However, it is important to clearly establisha common definition of the term when discussing thisquality with individuals from different sectors of theagricultural industry. Growers, researchers, plantbreeders, and seed sellers may have slightly differentunderstandings of the term. In addition, the word *tolerance,*which has a slightly different meaning, is sometimesused interchangeably with resistance, resulting insome confusion. By definition, tolerant plants canendure severe disease without suffering significantlosses in quality or yield; however, these tolerant plantsdo not significantly inhibit the pathogen's activity, anddisease symptoms may be clearly evident. Resistantplants usually suppress the pathogen in some fashion.There are some distinct advantages to planting disease-resistant plant cultivars. Such selections are completelynon-disruptive to the environment, and in facttheir use enables growers to reduce and in some caseseliminate the application of chemicals used forpathogen control. The use of cultivars resistant to onedisease is compatible with disease management stepstaken to control other diseases. A final advantage is thatfor some host-pathogen systems the stability of resistance is long lasting and the cultivars can remainresistant for many years.

There are some disadvantages to the use of resistantcultivars. The greatest shortcoming is that resistance isnot available for all diseases on all crops. For several ofthe most damaging plant diseases, such as tomato lateblight (*Phytophthorainfestans*) and white rot (*Sclerotiumcepivorum*) of Alliums, acceptable no resistant cultivarsare yet available. In addition, commercial seed companiesand plant breeders rarely invest in efforts to developresistant cultivars for specialty or minor crops.Hence there will be specialty commodities, many ofwhich are popular choices for organic producers thatwill continue to lack resistance to their disease problems.

Applying Control Materials

Disease control using microorganisms (biocontrol)or chemical by-products made by microorganisms isgenerating a good deal of interest. However, the historyof successful biological control of plant diseases is notencouraging. Very few effective, economically feasiblebiological control materials are commercially available.Much research and development remains to be done.For the best results possible with any of these materials,appropriate application technique (proper equipment,spray volume, and plant coverage) and timing are essential. Most materials do not perform well if thedisease is established, so applications should be madeprior to extensive infection. Before applying a product,a grower should confirm that the material is approvedfor use in organic production. Consult product labels,UC Cooperative Extension farm advisors, pest controladvisers, and your local Agricultural Commissioner'sOffice for product use information and restrictions.

References

A.H.C. van Bruggen and M.R. Finckh, Plant Diseases and Management Approaches in Organic Farming Systems, Annual Review of Phytopathology, 10.1146/annurev-phyto-080615-100123, 54, 1, (25-54), (2016).

Aggarwal A, Kadian N, Tanwar A, Yadav A, Gupta KK. Role of arbuscularmycorrhizal fungi (AMF) in global sustainable development. Journal of Applied and Natural Science, 2011. 3(2):340–351.

Axel Mie, HelleRaun Andersen, Stefan Gunnarsson, Johannes Kahl, Emmanuelle Kesse-Guyot, EwaRembiałkowska, GianlucaQuaglio and Philippe Grandjean, Human health implications of organic food and organic agriculture: a comprehensive review, Environmental Health, 10.1186/s12940-017-0315-4, 16, 1, (2017).

DavideSpadaro and Maria LodovicaGullino, Sustainable Management of Plant Diseases, Innovations in Sustainable Agriculture, 10.1007/978-3-030-23169-9_11, (337-359), (2019).

Dun-chun HE, Jia-sui ZHAN and Lian-hui XIE, Problems, challenges and future of plant disease management: from an ecological point of view, Journal of Integrative Agriculture, 10.1016/S2095-3119(15)61300-4, 15, 4, (705-715), (2016).

10

Types of Water Sources for Irrigation in Agriculture

A.K. Sinha, Vinod Gupta, Neerja Sharma and Saurav Gupta

KVK Samba, SKUAST-Jammu, Jammu and Kashmir, India

Irrigation

Irrigation means the watering of land to make it ready for agriculture. It is the process of application of water to crops through artificial channels to grow them. Water is vital for the growth of plants. And, there can be no plants or crops if they do not have access to water in some form. It is, thus, crucial to supply water to crops and plants in time as per their need. The supply of water to plants comes from various water resources. For example,

- Wells
- Ponds
- Lakes
- Canals
- Dams and Reservoirs

Importance of Irrigation

Irrigation is necessary for agriculture and farming.

1. Firstly, it enables growth and photosynthesis in plants. Plants absorb minerals and nutrients from the soil via their roots. These minerals dissolve in the water present in the soil. Then the water transports these nutrients to all parts of the plant. In this way, it enables growth and photosynthesis.
2. Secondly, it provides the moisture that is crucial during the germination phase of the plant's life cycle.

3. Thirdly, it helps increase soil fertility by adding moisture to it. It also makes the land easier to plough.
4. Lastly, it increase the yield from the farm.

Traditional Irrigation Methods

These irrigation systems were used in earlier years. However, even today some small farms in rural areas adopt these. Although they are cheaper than modern methods, they are not as efficient. Because they need human or animal labour to function. Some of these systems are,

1. **MOAT:** Moat is also called the pulley system, it involves pulling up water from a well or other such source to irrigate the land. This is a time consuming and labour intensive process, but it is very cost-efficient. Also, using a moat avoids wastages of water.
2. **Chain pump:** A chain pump consists of two large wheels connected by a chain. There are buckets attached to the chain. Further, one part of the chain dips into the water source. As the wheels turns, the bucket picks up water. The chain later lifts them to the upper wheel where the water gets deposited into a source. The empty bucket gets carried back down.
3. **Dhekli :** It is a process of drawing water from a well or such similar source. Here we tie a rope and bucket to a pole. At the other end, we tie a heavy stick or any other object as a counterbalance. And we use this pole to draw up water.
4. **Rahat :** Rahat uses animal labour. Above the well, we tie a large wheel. An ox or cow would turn the wheel to draw the water from the well.

Five highly effective methods of irrigation

These are more efficient irrigation systems that were invented in recent decades. These help us use water economically without wastage for agriculture.

1. Drip Irrigation system
2. Sprinkler system
3. Surface irrigation
4. Basin irrigation
5. Furrow irrigation

1. Drip Irrigation system

Drip irrigation is the most used irrigation system these days. Drip irrigation can be defined as the method in which water drips slowly via a pipe system to the roots of the plants either from above or below the soil surface. In the drip irrigation, we lay plastic pipes in rows near the crops or plants. These pipes have holes in them. The water seeps from these holes drop by drop, hence the name drip irrigation. Drip irrigation is one of the most efficient irrigation methods as it reduces water wastage in agriculture.

2. Sprinkler system

Sprinkler systems mimic the phenomenon of rain. In sprinkler systems, Water is pumped using a pipe system and then sprayed through sprinkler heads. The pipes carry water to central locations on the farm. The sprinkler method is one of the most efficient irrigation methods to irrigate the uneven land for agriculture. In addition, sprinkler systems provide the best coverage regardless of the size of the farm.

3. Surface Irrigation

Surface irrigation has been practiced and followed for many years now.It can be defined as a group of techniques where water is distributed over the surface of the soil gravity. In this type of irrigation, either the field is flooded (this is known as Basin irrigation)or the water is fed into small channels(this is known as furrow irrigation).

4. Basin Irrigation

Basin irrigation method is primarily used for crops that stand in water for more extended periods, flat lands where rice is grown or in terraces on hills. In Basin irrigation flat areas of land are surrounded by low bunds. These bunds block the water and prevent it from entering the adjacent fields. Trees can also be grown using basin irrigation method. Basin irrigation is suitable pastures, citrus, banana and to some extent tobacco. This method cannot be used for crops that cannot stand waterlogged like potatoes, beetroot and carrots. The type of crop grown determines the soil suitable for basin irrigation. Basin irrigation can be constructed on a flat surface,the easier it is to build basins, sloping land.

5. Furrow Irrigation

The application in which a small channel carry water in between the crop rows and down the slope is known as Furrow irrigation.Furrow irrigation is preferable to row crops and the ones that cannot thrive water logging. Only flower, maize, sunflower, sugarcane and soybean can be irrigated via furrow irrigation.In this particular method of irrigation water flows from the field channel into the furrows by opening up the bank of the channel or by siphons or spiles.Furrows must determine the slope, type of soil, size of the stream, irrigation depth and field length.It should be done on flat or gentle slopes, if done on undulating land, furrow irrigation should follow the land contouring method.

References

Agrawal R. and Jain R.C. (1996). "Forecast of sugarcane yield using eye estimate along with plant characters". Biometrical Journal, 38(5):731-39

Babaria, N.B., M.S. Solanki, A.V. Ardeshana, and V.G. Barad (2005), 'Quality of Underground Irrigation Waters of Saurashtra Region', in N.C. Patel, A.R. Subbaiah, K.C. Patel, and J.C. Nandasana (eds) Sustainable Management of Water Resources, Himanshu Publication, New Delhi, pp. 144–5.

Briscoe, John and R.P.S. Malik (2006), India's Water Economy: Bracing for a Turbulent Future. Th e World Bank, Oxford University Press, New Delhi

Gandhi, Vasant and Ashutosh Roy (2009), 'Th e Performance of Ground-water Institutions in India: Empirical Study', in Lin Crase and Vasant P. Gandhi (eds), Reforming Institutions in Water Resource Management: Policy and Performance for Sustainable Development, Earthscan, London, pp. 237–59.

Gandhi, Vasant and Suresh Sharma (2009), 'Th e Performance of Rain-water Harvesting Institutions in India: Empirical Study', in Lin Crase and Vasant P. Gandhi (eds), Reforming Institutions in Water Resource Management: Policy and Performance for Sustainable Development, Earthscan, London, pp. 260–80.

11

Importance of Farm Mechanization in Floriculture

A.K. Sinha, Vinod Gupta, Saurav Gupta and Neerja Sharma

KVK Samba, SKUAST-Jammu, Jammu and Kashmir, India

Introduction

Mechanized agriculture is the process of using agricultural machinery to mechanize the work of agriculture, greatly increasing farm worker productivity. The effective mechanization contributes to increase production in two major ways: firstly the timeliness of operation and secondly the good quality of work. The requirement of power for certain operations like seedbed preparation, cultivation and harvesting becomes so great that the existing human and animal power in the country appears to be inadequate. As a result, the operations are either partially done or sometimes completely neglected, resulting in low yield due to poor growth or untimely harvesting or both.

Scope of Mechanization

It is quite true that the Indian farmers have the lowest earnings per capita because of the low yield per hectare they get from their holdings. One of the few important means of increasing farm production per hectare is to mechanize it. Mechanization in India may have to be done at various levels. Broadly, it can be done in three different ways: I. By introducing the improved agricultural implements on small size holdings to be operated by bullocks. II. By using the small tractors, tractor-drawn machines and power tillers on medium holdings to supplement existing sources. III. By using the large size tractors and machines on the remaining holdings to supplement animal power source. But many people are of the opinion that Indian agriculture cannot be fully mechanized. Only the improved animal-drawn implements should be introduced. It is felt that

1. There is a surplus of agricultural labour in India.
2. There are enough draft animals available in the country to do the farm work effectively.

3. The size of farm holdings of the majority of the Indian farmers is too small to justify the use of a tractor on their farms.
4. The investing capacity, of the farmers is too poor to buy a tractor and tractor-drawn implements.
5. The technical know-how of the people in the country is low.
6. In the absence of suitable farm road system, the tractor and tractor-drawn machines cannot be effectively utilized under the present conditions.
7. It will not be possible to increase the yield by using mechanical power.
8. Mechanization will not result in lowering the cost of production.
9. It will not be possible to mechanize every bit of farm operation.
10. A large labour force will get displaced from agriculture.

History indicates that the development in farm mechanization is very closely related to the shortage of human labour and industrial development in the country. Farmers of India like their counterparts in other countries are interested to improve their income, life style and general well being. They see mechanization as a means for achieving their objectives. In spite of the limitations with which they exist, their performance has been note worthy. There is a positive correlation between application of improved technologies and the land productivity.

Impact of Mechanization on Productivity

The technical know-how of the people in the country is quite satisfactory. It always develops with the opportunities and experience. The repair, maintenance and overhaul facilities for tractors and other machines are expanding, even in the small towns, with the expansion of rural electrification in the country. Drudgery and physical exertion are typical of much Indian agriculture today. It scares away men of intelligence and ability from agriculture. All people would favour minimizing this to a considerable extent. The development of improved riding type animal drawn machines can improve the present condition, but they cannot be a substitute for the tractor drawn machines. By using mechanical power, man will be able to control larger areas and as such his family members will get more free time. More power is essential in carrying out operations effectively at the right time and for changing the attitudes and uplifting the social status and dignity of those who work in agriculture. There are three ways in which progress can be made towards increased power, all of which must be worked on simultaneously in combination with integrated and matched implements.

1. By improving bullock harness and hitches;
2. By developing and introducing small tractors;

3. By increasing the number of large horsepower tractors.

These machines will be helpful in providing power efficiently for good seedbed preparation, which is quite essential for maximizing the germination of the seed and seedling growth. In addition to this, the mechanization of the following fields of agriculture- is equally essential.

- Shaping and levelling of farm fields for getting even distribution and depth of irrigation water;
- Development of planting and fertilizing machines to place the seed more precisely in rows and to place the fertilizer correctly with respect to seed or plant;
- Spraying and dusting machinery to apply chemicals for weed and pest control;

After production operations such as harvesting, threshing, winnowing and drying.

State of Mechanization

Agricultural system all over the world has undergone changes in terms of cropping system, type of power sources used and application of inputs to achieve high level of productivities. Even in India, mechanization of agriculture has advanced considerably. In certain region, the level of mechanization has gone far ahead of the average level in the country. Human and animal power sources are no longer the predominant sources on Indian farms. The annual addition of tractor population is more than one hundred thousand tractors. Similarly about the hundred thousand pump sets are being installed on Indian farms annually. On the basis of annually critical review of the mechanization position, one observes that the shortage of labour and high labour wages are the factors which strongly propel mechanization. Consequently, the more labour intensive operations, such as pumping of irrigation water, land preparation and threshing are the first operations, which are mechanized. Large amount labour or draft power, which can be replaced through machines, provides a strong incentive to mechanize. The farm operations, which can be categorized as:

- Highly power intensive operations,
- Intermediate power and control intensive operations,
- Highly control intensive operations.
    - Water Pumping
    - Tillage
    - Direct Seeding

- Transplanting
- Weeding
- Plant protection
- Harvesting
- Threshing
- Milling
- Transport

Social Consideration to Farm Mechanization

The steps towards development of an appropriate agricultural technique in India are not working with the motto of saving labour but of improving and increasing agricultural production. Most people believe that India is burdened with surplus labour and that there are far too many draft animals for the cultivation of available farmland. Above all, there is an unfounded fear that farm mechanisation will precipitate widespread rural unemployment as agriculture contributes over two-thirds of all jobs in India. The fact is however, that India's increased food requirements must be met through increased productivity of the land from higher yields and more multiple cropping which would require additional labour for carrying out different farm operations. From the energy application point of view, the Indian agriculture is in the transition from stage 1 (human power) and stage 2 (animal power) to stage 3 and 4 (power tiller or four wheel tractor). However, animal power will co-exist with mechanical power in the country. Animal drawn machines must be provided with tiding arrangements so that walking behind the machine can be avoided. The following can help in increasing the agricultural production in the country:

- Agriculture must continue to productively employ as many or more laborers per cultivated hectare in the foreseeable future.
- The small & scattered land holdings and the poor economic conditions of the average farmer restrict the use of units designed for large-scale mechanization. The power tillers and related equipment, therefore, find a greater scope to be used on such farms. The power availability on the farm should be increased by 2.5 times more.
- Cropping intensity should be increased to 200 to 250 per cent level from the present level of 100 to I80 per cent.
- Due to the fragmentation of lands, the number of holdings has gone up and therefore, appropriate size of farm machines have to the introduced for the tillage, planting transplanting and harvesting of crops. Custom hiring of

machines will have large scope in future.

- Custom hiring of farm equipment should be encouraged as it has resulted to the increased power availability on the farms and in turn increased land productivity.
- Post harvest technology deserves special attention on Indian farms.

References

Anonymous 2013. State of Indian Agriculture 2012-13. Directorate of Economics and Statistics, Department of Agriculture and Cooperation, Ministry of Agriculture, New Delhi http://164.100.47.132/paper laid files/AGRICULTURE/State of Indian Agriculture 2012-13.pdf, 221 p.

Henten E.J.V., (2006). Greenhouse Mechanization: State of the Art and Future Perspective, Acta Hort. 710.

Kulkarni S. D. 2005. Country Report–India: Agricultural Mechanization – Present Scenario and Perspective. Presented in Fourth Session of the Technical Committee of APCAEM held at New Delhi, India during November 21-24, 2005

Mehta C. R., Pajnoo R. K. 2013. Role of Japan in Promotion of Agricultural Mechanization in India. Agri. Mechanization in Asia, America and Latin America, 44: 15-17

Peterson D.L., (2005). Harvest mechanization progress and prospects for fresh market quality deciduous tree fruits. Hort. Techn. 15(1).

Peterson D.L., Whiting M.D. and Wolford S.D., (2003). Fresh market quality tree fruit harvester Part I: Sweet cherry. Appl. Eng. Agr. 19(5):539-543.

Poole W.D., (1968). The impact of mechanization of horticultural crop production in Louisiana on teaching, NACTA Journal, 82-84.

12

Floriculture Its Prospects and Scope

Neerja Sharma, Vinod Gupta, Saurav Gupta Rakesh Sharma and A.K. Sinha

KVK Samba,SKUAST-Jammu, Jammu and Kashmir, India

Introduction

Floriculture or flower farming is the study of growing and marketing flowers and foliage plants. Floriculture includes cultivation of flowering and ornamental plants for direct sale or for use as raw materials in cosmetic and perfume industry and in the pharmaceutical sector. It also includes production of planting materials through seeds, cuttings, budding and grafting. In simpler terms floriculture can be defined as the art and knowledge of growing flowers to perfection. The persons associated with this field are called floriculturists.

Worldwide more than 140 countries are involved in commercial Floriculture. The leading flower producing country in the world is Netherlands and Germany is the biggest importer of flowers. Countries involved in the import of flowers are Netherlands, Germany, France, Italy and Japan while those involved in export are Colombia, Israel, Spain and Kenya. USA and Japan continue to be the highest consumers.

Floriculture in India

Floriculture is an age old farming activity in India having immense potential for generating gainful self-employment among small and marginal farmers. In the recent years it has emerged as a profitable agri-business in India and worldwide as improved standards of living and growing consciousness among the citizens across the globe to live in environment friendly atmosphere has led to an increase in the demand of floriculture products in the developed as well as in the developing countries worldwide. The production and trade of floriculture has increased consistently over the last 10 years. In India, Floriculture industry comprises flower trade, production of nursery plants and potted plants, seed and bulb production, micro propagation and extraction of essential oils. Though the annual domestic demand for the flowers is growing at a rate of over 25% and international demand at around Rs 90,000 crore India's share in international market of flowers is negligible. However, India is having a better scope in the future as there is a shift in trend towards tropical flowers and this can be gainfully exploited by country like India with high amount of diversity in indigenous flora.

After liberalization the Government of India identified floriculture as a sunrise industry and accorded it 100 percent export oriented status. The liberalization of industrial and trade policies paved the way for the development of export oriented production of cut flowers. The new seed policy has already made it feasible to import planting material of international varieties. Floriculture products mainly consist of cut flowers, pot plants, cut foliage, seeds bulbs, tubers, rooted cuttings and dried flowers or leaves. The important floricultural crops in the international cut flower trade are rose, carnation, chrysanthemum, gerbera, gladiolus, orchids, anthurium, tulip and lilies.

According to statistics indicated in the Handbook on Horticulture Statistics 2014, the total area under flower crops in 2012-13 was 232.70 thousand hectares. Total area under floriculture in India is second largest in the world and only next to China. Production of flowers was estimated to be 1729.2 MT of loose flowers and 76731.9 million (numbers) of cut flowers in 2012-13. Fresh and Dried cut flowers dominate floriculture exports from India.Among states, Karnataka is the leader in floriculture with about 29,700 hectares under floriculture cultivation. Other major flower growing states are TamilNadu and Andhra Pradesh in the South, West Bengal in the East, Maharashtra in the West and Rajasthan, Delhi and Haryana in the North.

The expert committee set up by Govt. of India for promotion of export oriented floriculture units has identified Bangalore, Pune, New Delhi and Hyderabad as the major areas suitable for such activity especially for cut flowers. Of the four zones identified as potential centers for flower production namely Bangalore, Hyderabad, Pune and New Delhi, the area around Bangalore and Pune have got the advantage of ideal climatic conditions where the temperature ranges between 15 to 30°C. In view of this, the units established in these locations do not require either cooling or heating system. As a result maximum number of units has been established in these locations. There are more than 300 export oriented units in India. APEDA (Agricultural and Processed Food Products Export Development Authority) is the registering authority for such units.

Marketing

In India Marketing of cut flowers is much unorganized. In most of the Indian cities flowers are brought to wholesale markets, which mostly operate in open yards. From here the flowers are distributed to the local retail outlets which more often than not operate in the open on-road sides, with different flowers arranged in large buckets. In the metropolitan cities, however, there are some good florist show rooms, where flowers are kept under controlled temperature conditions, with considerable attention to value added service.The government is now investing in setting up of auction platforms, as well as organized florist shops with better storage facilities to prolong shelf life. The packaging and transportation of flowers from the farms to the retail markets at present is very unscientific. The flowers, depending on the kind, are packed in gunny bags, bamboo baskets, simple cartons or just wrapped in old newspapers and transported to markets by road, rail or by air. However, the government has provided some assistance for buying refrigerated cargos and built up a large number of export oriented units with excellent facilities of pre-cooling chambers, cold stores and reefer vans.

According to a study titled, 'Indian Floriculture Industry: The Way Ahead' released by the apex industry body ASSOCHAM, India's floriculture industry is growing at a compounded annual growth rate of about 30%, and is likely to cross Rs 8,000 crore mark by 2015. Currently, the floriculture industry in India is poised at about Rs 3,700 crore with a share of a meager 0.61% in the global floriculture industry which is likely to reach 0.89% by 2015.

Export Constraints

In spite of an abundant and varied production base, India's export of floricultural product is not encouraging. The low performance is attributed to many constraints like non-availability of air space in major airlines. The Indian floriculture industry is facing with a number of challenges mainly related to trade environment, infrastructure and marketing issues such as high import tariff, low availability of perishable carriers, higher freight rates and inadequate refrigerated and transport facilities. At the production level the industry is faced with challenges mostly related to availability of basic inputs including quality seeds and planting materials, efficient irrigation system and skilled manpower. In order to overcome these problems, steps must be taken to reduce import duty on planting material and equipment, reduce airfreight to a reasonable level, provide sufficient cargo space in major airlines and to establish model nurseries for supplying genuine planting material. Training centers should be established for training the personnel in floriculture and allied areas. Exporters should plan and monitor effective quality control measures right from production to post harvesting,storage, and transportation.

Government Programmes and Policies

Department of Agriculture and Cooperation under the Ministry of Agriculture is the nodal organization responsible for development of the floriculture sector. It is responsible for formulation and implementation of national policies and programmes aimed at achieving rapid agricultural growth through optimum utilization of land, water, soil and plant resources of the country. Production of cut flowers for exports is also a thrust area for support. The Agricultural and Processed Food Products Export Development Authority (APEDA), the nodal organization for promotion of Agri-exports including flowers, has introduced several schemes for promoting floriculture exports from the country. These relate to development of infrastructure, packaging, market development, subsidy on airfreight for export of cut flowers and tissue-cultured plants, database up-gradation etc. The 100% Export Oriented Units are also given benefits like duty free imports of capital goods. Import duties have also been reduced on cut flowers, flower seeds, tissue-cultured plants, etc. Setting up of walk in type cold storage has been allowed at the International airports for storage of export produce.

Initiatives have also been launched for the benefit of exporters by providing cold storage and cargo handling facility for perishable products at various international airports. Direct subsidy up to 50 percent is also available in cold storage units. Besides, subsidy is also provided by APEDA on improved packaging materials to promote their use. To attract entrepreneurship in floriculture sector, NABARD is providing financial assistance to hi-tech units at reasonable interest rates.Several schemes have been initiated by the Government for promotion and development of the floriculture sector including "Integrated Development of Commercial Floriculture" which aims at improvement in production and productivity of traditional as well as cut flowers through availability of quality planting material, production of off season and quality flowers through protected cultivation, improvement in post harvest handling of flowers and training persons for a scientific floriculture. Many state governments have set up separate departments for promotion of floriculture in their respective states.

Research work on floriculture is being carried out at several research institutions under the Indian Council of Agricultural Research and Council of Scientific and Industrial Research, in the horticulture/floriculture departments of State Agricultural Universities and under the All India Coordinated Floriculture Improvement Project with a network of about twenty (20) centers. The key focus areas are crop improvement, standardization of agro-techniques including improved propagation methods, plant protection and post harvest management. In recent years, however, technologies for protected cultivation and tissue culture for mass propagation have also received attention. A large number of promising varieties of cut flowers have been developed. All these efforts indicate the government's commitment for improving the sector and creating a positive environment for entrepreneurship development in the field.

Demand and Supply

The demand for flowers is seasonal as it is in most countries. The demand for flowers has two components: a steady component and a seasonal component. The factors which influence the demand are to some extent different for traditional and modern flowers.

(i) Traditional Flowers

The steady demand for traditional flowers comes from the use of flowers for religious purposes, decoration of homes and for making garlands and wreaths. This demand is particularly strong in Kerala, Karnataka, Tamil Nadu, Odisha and West Bengal, as the use of flowers for above mentioned purposes is part of their local culture. The bulk of seasonal demand comes from festivals and marriages. The demand is generally for specific flowers.

(ii) Modern Flowers

The bulk of the steady demand for modern flowers comes from institutions like hotels, guest houses and marriage gardens. The demand is concentrated in urban areas. With increasing modernization and globalization the demand for modern flowers from the individual consumers is likely to grow enormously as the trend of "say it with flowers" is increasing and the occasions which call for flower giving will continue to present themselves. Although there is an increasing demand for modern flowers from individuals, institutions continue to be the dominant buyers in the market. The price of these flowers also depends on their demand and varies accordingly.

Green House Technology for Flower Production

In present scenario of increasing demand for cut flowers protected cultivation in green houses is the best alternative for using land and other resources more efficiently. In protected environment suitable environmental conditions for optimum plant growth are provided which ultimately provide quality products. Green House is made up of glass or plastic film, which allows the solar radiations to pass through but traps the thermal radiations emitted by plants inside and thereby provide favourable climatic conditions for plant growth. It is also used for controlling temperature, humidity and light intensity inside. On the basis of basic material used, building cost and technology used, green houses can be of three types:

1. **Low-cost greenhouse:** The low-cost green house is made of polythene sheet of 700 gauge supported on bamboos with twines and nails. Its size depends on the purpose of its utilization and availability of space. The temperature within greenhouse increases by 6-10^{0}C more than outside.

2. **Medium-cost greenhouse:** With a slightly higher cost greenhouse can be framed with GI pipe of 15 mm bore. This greenhouse has a covering of UV -stabilized polythene of 800 gauge. The exhaust fans are used for ventilation which are thermostatically controlled. Cooling pad is used for humidifying the air entering the chamber. The greenhouse frame and glazing material have a life span of about 20 years and 2 years respectively.
3. **Hi-tech greenhouse:** In this type of green house the temperature, humidity and light are automatically controlled according to specific plant needs. These are indicated through sensor or signal-receiver. Sensor measures the variables, compare the measurement to a standard value and finally recommend to run the corresponding device. Temperature control system consists of temperature sensor heating/cooling mechanism and thermostat operated fan. Similarly, relative humidity is sensed through optical tagging devices. Boiler operation, irrigation and misting systems are operated under pressure sensing system. This modern structure is highly expensive, requiring qualified operators, maintenance, care and precautions. However, these provide best conditions for export quality cut flowers and are presently used by large number of export units.

Floriculture has emerged as an important agribusiness, providing employment opportunities and entrepreneurship in both urban and rural areas. National Horticulture Board helps one to establish a flower business. Agricultural and Processed Food Products Export Development Authority helps entrepreneurs with cold storage facilities and freight subsidies. It has been found that Commercial Floriculture has higher potential per unit area than most of the field crops and therefore a lucrative business. During the last decade there has been a thrust on export of cut flowers. The export surplus has found its way into the local market influencing people in cities to purchase and use flowers in their daily lives. Floriculture thus, offers a great opportunity to farmers in terms of income generation and empowerment. Small and marginal farmers may also use every inch of their land for raising the flower and foliage crops. Floriculture also offers careers in production, marketing, export and research. One can find employment in the floriculture industry as a farm manager, plantation expert, supervisor or project coordinator. Besides, one can work as consultant or landscape architect with proper training. In addition, floriculture also provides career opportunities in service sector which include such jobs as floral designers, landscape designers, landscape architects and horticultural therapists. Research and teaching are some other avenues of employment in the field.

References

Agoramoorthy, G. & Hsu, M. J. (2012) Impact of floriculture development enhances livelihood of India's rural women, Journal of Agriculture and Rural Development in the Tropics and Subtropics, 113(1), pp: 69-76.

Mathur, R. and Pachpande, P. (2013) Floriculture prospects and opportunities, International E Journal Management and IT, 13(1), pp: 1-10

Niranjan, S. K. D. F. and Gunasena, H. P. M. (2006) Floriculture sector development programme: small and medium scale entrepreneurs in Sri Lanka, Sri Lanka Council for Agricultural Research Policy, pp: 1-49

Noreen, S. (2011) Role of microfinance in empowerment of female population of Bahawalpur District, International Conference on Economics and Finance Research IPEDR,4 (1), pp: 23-37

Rani, S. (2014 Entrepreneurial empowerment of women through Self- Help group, International Journal of Research in IT & Management, 4(2), pp: 27-35

13

Organic Farming in India A Vision Towards a Healthy Nation

Vinod Gupta, Neerja Sharma, Saurav Gupta and A.K. Sinha

KVK, Samba, SKUAST-Jammu, Jammu and Kashmir, India

Introduction

Indian economy depends on agriculture and the agriculture sector in India contributes about17 percent of country's GDP and provides employment to approximately two third of the population (United Nation Development Programme Fact sheet 2011). India is the second most populous country in the world. With the increasing population, the cultivable land resource is shrinking day by day. To meet the food, fibre, fuel, fodder and other needs of the growing population, the productivity of agricultural land and soil health needs to be improved. Green Revolution ushered through use of modern agriculture technologies such as greater use of synthetic agrochemicals like fertilizers and pesticides, adoption of nutrient-responsive, high-yielding varieties of crops, greater exploitation of irrigation potentials, etc. undoubtedly has increased production and labour efficiency in the post-independence era and made our countries self-sufficient in food production. But indiscriminate and excessive use of chemicals during this period has put forth a question mark on sustainability of agriculture in the long run and concerns have been raised time and again over its adverse effects on soil productivity and environment. Some of the adverse effects are soil erosion, depletion of organic matter in soil, nutrient deficiencies, low water availability, salinization, fertilizer and pesticide contamination of food and water bodies and erosion of bio-diversity calling attention for sustainable production technology which will address soil health, human health and environmental health and eco-friendly agriculture. Besides in the present era of global warming and climate change, the face of agriculture has to be more environments friendly, hence the main emphasis should be for development of production technologies which are sustainable in long run. Therefore, in sustaining agricultural production against the finite natural resource base, the demands have shifted from the "resource degrading" chemical agriculture to a "resource protective" biological or organic agriculture. Organic farming appears to be one of the options for sustainability. During the last two decades, 'Organic Agriculture' has emerged as a dynamic 'Alternate Farming System'. The primarily goal of organic farming is to optimize the health and productivity of interdependent communities of soil, plants, animals and people. The primarily goal of organic vegetable production is to optimize the health and productivity of interdependent communities of soil, plants, animals and people.

Basic Concept and Principles of Organic Farming

The concept of organic farming is not clear to many concerns. Many people consider that traditional agriculture, sustainable agriculture, JaivikKrishietc, as organic farming. Some people are of the idea that the use of organic manures and natural methods of plant protection instead of using synthetic fertilizers/ pesticides is organic farming.

The basic concepts behind Organic Farming are

- It concentrates on building up the biological fertility of the soil so that thecrops take the nutrients they need from the steady turnover within the soil nutrients produced in this way are released in harmony with the needs of the plants.
- Control of pests, diseases, and weeds is achieved largely by the development of an ecological balance within the system and by the use of bio-pesticides and various cultural techniques such as crop rotation, mixed cropping, and cultivation.
- Organic farmers recycle all wastes and manures within a farm but the export of the products from the farm results in a steady drain of nutrients.
- In a situation, where conservation of energy and resources is considered to be important, community or country would make every effort to recycles to all urban and industrial wastes back to agriculture and thus the system would be only being small inputs of new resources to "top up" soil fertility.

According to IFOAM, organic agriculture is guided by following four principles:

Principle of health: Organic agriculture should sustain and enhance the health of soil, plant, animal, human and planet as one and indivisible. This principle points out that the health of individuals and communities cannot be separated from the health of ecosystems - healthy soils produce healthy crops that foster the health of animals and people.

Principle of ecology: Organic agriculture should be based on living ecological systems and cycles, work with them, emulate them and help sustain them. Thisprinciple roots organic agriculture within living ecological systems. It states that production is to be based on ecological processes, and recycling.

Principle of fairness: Organic agriculture should build on relationships that ensure fairness with regard to the common environment and life opportunities. Fairness is characterized by equity, respect, justice and stewardship of the shared world, both among people and in their relations to other living beings.

These principles of organic farming encourage:

- To work as much as possible within a closed system, and draw upon local resources:
- To maintain the long-term fertility of soils.
- To avoid all forms of pollution that may result from agricultural techniques.
- To produce foodstuffs of high nutritional quality and sufficient quantity.
- To reduce the use of fossil energy in agricultural practice to a minimum.
- To give livestock conditions of life that confirm to their physiological need.

The organic agriculture is called in different names by different people, however, the basic concept and philosophy remains the same. A number of terminologies are available as mentioned below:

i. Organic farming: It is a holistic production management system that avoids the use of synthetic fertilizers and pesticides, minimizes pollution of air, soil and water, and optimizes the health and productivity of independent communities of life, plants, animals andpeople. It uses the organic amendments to supply nutrition to the soil and crop.

ii) Bulky organic manures (viz. FYM, vermin-compost etc.)

iii) Green manure (viz. dhaincha, sunnhemp etc.)

iv) Concentrated organic manures (viz. groundnut cakes, neem cake etc.).

Natural farming: It was developed in Japan during 1930 by Mokichi Okada who later formedthe "Mokichi Okada Association (MOA)". Natural farming is at par with organic farming in many ways but gives special emphasis on soil health through composts and use of microbial preparations.

Ecological agriculture: It is labour intensive system which is based on techniques of crop cultivation.It involves all attention to promote renewable source of energy (draught animalpower), electric energy from garbage disposal and biogas from organic wastes. It also involves water use efficiency through conjunctive use of rain, tank, underground well and river water. Practices which improve crop productivity (genetic and agronomic) i.e. hybrid vigour, gene pyramiding, multiple cropping, Integrated Plant Nutrient Management (IPNM) and Integrated Pest Management (IPM) are the essential components of ecological farming.

Agnihotra: Agnihotra is a process of purification of the atmosphere as a cumulative effect of various scientific principles harnessed to give rise to an unparalleled purifying and the process of Agnihotra consists of making two offerings to the fire exactly at the time of sunrise and sunset along with the chanting of two smallSanskrit mantras. Agnihotra balances the cycle of nature and nourishes the humanlife. It cleanses the negative effects of pollution.

Rishi Krishi (Rishi Kheti): This system was evolved by Mr. Mohan Shankar Deshpande in Maharashtra.In this system, the following aids are used to maintain soil fertility and crop yield.

a) **Angara**: BhomiSanskar is being performed to make the soil fertile wherein 15 kg rhizosphere soil of banyan tree (Ficusbenghalensis) is broadcasted over an acre of land. It has lot of earthworms and other beneficial microbes, which improves thesoil fertility.

b) **AmritPani**: AmritPani is prepared by mixing 20 kg cow dung, 1⁄2 kg honey, 1⁄4 kg ghee. All ingredients are mixed and kept overnight. This is used to treat the seeds field and plants(BeejSanskar, JalaSanskar and VanaspatiSanskar).

c) **PanchyaGavya:** It basically consists of five products from cow i.e. dung (5 kg),urine (5 litres), milk (3 litres), curd (3 litres) and ghee (1 kg). These are mixed with sugarcane juice, tender coconut water, ripe banana and toddy and incubated for 15days. The mixture is stirred daily for proper mixing and fermentation. Mixture is diluted in 1:10 ratio with water, filtered and sprayed on the crop.

Homeopathy farming: It is a kind of farming with homeopathic aids i.e. Homeo nutrients and Homeopathic plant protection measures.

Biodynamic agriculture or Vedic Kheti: Biodynamic farming is defined as working with energies, which create and maintain life. It involves certain principles and practices for healthy soil, healthy plant and healthful food for human beings and feed for animals. In the system, energies from cosmos, earth, cow and plants are systematically and synergistically harnessed. It is based on the knowledge that soil, plants, animals and men work together in one agricultural cycle.The prepared substances are numbered 500 through 508, where the first two are used for preparing fields whereas the latter seven are used for making compost.

BD 500: It is a humus mixture prepared by filling the horn of a cow with cow manure and burying it in the ground (40–60 cm below the surface) in the autumn. It is left to decompose during the winter and recovered for use the following spring.

BD 501: It is prepared by stuffing crushed and powdered quartz into a horn of a cow and buried into the ground in spring and taken out in autumn. It can be mixed with 500 but usually prepared on its own (mixture of 1 tablespoon of quartz powder to 250 liters of water) The mixture is sprayed under very low pressure over the crop during the wet season, to prevent fungal diseases. It should be sprayed on an overcast day or early in the morning to prevent burning of the leaves.

BD 502: Yarrow blossoms (*Achilleamillefolium*) are stuffed into urinary bladders from Red Deer *(Cervuselaphus*), placed in the sun during summer, buried in earth during winter and retrieved in the spring.

BD 503: Chamomile blossoms (*Matricariarecutita*) are stuffed into small intestines from cattle buried in humus-rich earth in the autumn and retrieved in the spring.

BD 504: Stinging nettle (*Urticadioica*) plants in full bloom are stuffed together underground surrounded on all sides by peat for a year.

BD 505: Oak bark (*Quercusrobur*) is chopped in small pieces, placed inside the skull of a domesticated animal, surrounded by peat and buried in earth in a place where lots of rain water runs past.

BD 506: Dandelion flowers (*Taraxacumofficinale*) are stuffed into the mesentery of a cow and buried in earth during winter and retrieved in the spring.

BD 507: Valerian flowers (*Valerianaofficinalis*) are extracted into water. BD 508: Horsetail (Equisetum)

Astronomical planting calendar : In biodynamic farming, it is considers that there are astronomical influences on soil and plant development, for example, what phase of the moon is most appropriate for planting, cultivating or harvesting various kinds of crops. This aspect of biodynamic has been termed "astrological" in nature.

Characteristics of Organic Farming Systems

Management of Organic farming is focused on the whole farm system and its interactions with climate, environment, social as well as economic conditions, rather than considering the farm as comprises of individual enterprises. The key characteristics of Organic Farming include:

- Protecting the long-term fertility of soils by maintaining organic matter levels, soil biological activity and careful mechanical intervention.
- Nitrogen self-sufficiency through the use of legumes and biological nitrogen fixation, as well as effective recycling of organic materials, including crop residues and livestock wastes.
- Weed, disease and pests control relying primarily on crop rotation, natural predators, crop diversity, organic manuring, use of resistant varieties and limited thermal, biological and chemical intervention.

- Supplementing crop nutrients, where necessary, by using nutrient sources which are made available to the plants indirectly but the action of soil micro organisms and chemical reactions of the soil.
- The extensive management of livestock, paying full regards to their evolutionary adaptations behavioural needs, and animal welfare issues with respect to nutrition, housing, health, breading and rearing.

Objectives of Organic Farming

The primary goal of organic farming is to optimize the health and productivity of interdependent communities of soil, plant, animal and people

- To produce food of high nutritional quality in sufficient quantity
- To encourage biological cycles within farming systems by involving the use of microorganisms, soil flora & fauna, plants and animals
- To maintain and increase the long-term fertility of soil and biodiversity
- To use renewable resources in locally organized production systems
- To work as much as possible within a close system with regard to organic matter and nutrient elements and draw up on local resources
- To avoid all forms of pollution that may results from Agricultural techniques
- To reduce the use of fossil energy in agricultural practice to the minimum

Nutrient Source

Green manures

Green manuring refers to incorporation of live biomass into the soil in order to supply plant nutrients. When green and succulent tissues are used, decomposition of incorporated biomass is faster, thus the practice attained a name called 'green manuring'. The crops grown to be incorporated are called green manuring crops. Ideally, a green manure crop should be a fast growing, non-woody and short duration crop so that the practice of green manuring is practically feasible in a given cropping system. A large number of fast growing nitrogen fixing crops like dhaincha (Sesbania sp.), sunhemp and cowpea may be used as green manure which can fix atmospheric nitrogen to the extent of 60-100 kg/ha. Generally, dhaincha (Sesbaniaesculenta, S. rostrata) and sunhemp (Crotolariajuncia) are ploughed in the soil after about 6 to 8 weeks of sowing when adequate vegetative growth is attained. Use of green manure is highly beneficial for organic production and maintaining soil health Besides adding nutrients into the soil, green manures also improve the physical and microbial properties of the soil. The benefits arising out of green manuring practice are limited by factors such as:

- Quantity of green matter incorporated and actual nutrients contained in them
- Time taken for decomposition
- Favorable factors available in soil for decomposition
- Stage of incorporation
- Release pattern of nutrients

Farm Yard Manure

The term 'farm yard manure' is an expression to signify any manure prepared in the backyard using the farm waste, cattle urine and dung. Since very early ages, its use is prevalent in agriculture. Three principal methods of preparing FYM are standardized

1. Open pit method Allowing semi-aerobic decomposition of farm wastes in underground pit. This method gradually requires turning of the decomposition material frequently
2. Sealed pit method Allowing anaerobic decomposition of farm wastes in underground pit. This requires sealing the surface of the pit with dung slurry
3. Japanese method (vat method) Allowing aerobic decomposition of farm wastes in an above ground structures of stones/bricks or thatched walls upto height of 1 meter.

The response to applied FYM could be studied in four groups:

1. Improved crop yields
2. Improved physical properties of soil
3. Improved microbial activity
4. Residual benefits to succeeding crop

Vermi-compost

Vermiculture technology is an aspect involving the use of earthworms as versatile natural bioreactors for effective recycling of non-toxic organic wastes to the soil. They effectively harness the beneficial soil micro flora, destroy soil pathogens, and convert organic wastes into valuable products such as bio-fertilizers, bio-pesticides, vitamins, enzymes, antibiotics, and growth hormones.The term vermicompost refers to manure prepared out of large scale rearing of earthworms in an artificial or natural pit. Earthworms can be used in specifically designed pits or in any ordinary FYM pits. Frequently, a term vermiculture is also used to denote a culture of young earthworms which are let in any undigested compost pit or in an agricultural field directly. They reproduce and grow in number. The fecal waste from large number of earthworm is available as manure.

Concentrated organic manures (Oil cakes): Use of oil cakes as manures for increasing the crop yields is one of the conventional forms of organic farming. Before the use of fertilizers, the oil cakes were applied in granulated formulations to agricultural land. At present in the Indian scenario, use of oil cakes as a source of nutrients is largely limited because:

- Most of the edible oil cakes are fed to the cattle as concentrates. • Oil cakes usually supply limited amount of nutrients because of better oil extraction methods like solvent extraction as compared to traditional methods. • The extraction process is performed in oil industries which is located away from the agricultural fields thus increasing the cost of transportation per kg of nutrient in the oil cake. Nevertheless, the potentiality of oil cakes for being used as manures is evident in non- edible cakes (castor cake, linseed cake and neem cake) and oil bearing trees (Terminaliachebula, Madhucaindica, Jatrophacucras)

Liquid manures: Application of liquid manure (for soil enrichment) is essential to maintain the activity of microorganisms. 3-4 applications of liquid manure is essential for all types of crops. Liquid manures such as Jivamrutha, vermiwash, cow urine, Panchgavya compost tea and Biosol etc. are excellent growth promoters when used as foliar spray

Vermi-wash: Vermi-wash is prepared from the heavy population of earthworms reared in earthen pots or plastic drums. The extract contains major micronutrients, vitamins (such as B_{12}) and hormones (gibberellins) secreted by the earthworms. Earthworms produce bacteriostatic substances found in the vermi-wash can protect the bacterial infections.Vermi-wash can be sprayed on crops and trees for better growth, yield and quality

Jeevaamrit: Application ofJeevaamrit to soil improves the soil considerably. It also encourages microbial activity in the soil.

Preparation method: Jeevamrit is prepared by mixing 10 kg local cow dung with 10 litres cow urine, add 2 kg local jaggery, 2 kg pulse flour and handful of garden soil and the volume made upto 200 liters. Keep the drum in shade covering with wet gunny bag and stir the mixture clockwise thrice a day and incubate.

Application: Apply the mixture when the ground is wet for the plants. This seems to work wonders for the plants due to increased microbial activity by 3rd and 4th day. This is an excellent culture for enabling the exponential increase of beneficial microbes. The microbes are added through 2-3 handful of local soil.

Panchgavya: It is a cow excreta based indigenous nutrient solution. Panchagavya consists of products viz. cow dung, cow urine, milk, curd, jaggery, ghee, banana, tender coconut and water. When suitably mixed and used, these have miraculous effects.

Preparation Method: 1. 7 kg. cow dung and 1 kg. cow ghee is mixed thoroughly and kept for 3 days. After 3 days, 10 lt. cow urine and 10 lt. water is added, mixed and kept for 15 days with regular mixing both in morning and evening hours.. After 15 days the following ingredients are added and mixed Cow milk - 3 liters, Cow curd - 2 liters, Tender coconut water - 3 liters, Jaggery - 3 kg, Well ripened banana -12 nos. Panchagavya is ready after 30 days .In the investigation, 3% solution was found to be most effective compared to the higher and lower concentrations. 3 litres of Panchagavya to every 100 litres of water is ideal for all crops.

Biofertlizers: Biofertilizers either refer to a culture of microorganism applied to the soil/seed in order to encourage fixation of N or release of P or they include biomass of certain plants which contain large quantities of N as a result of their association with N fixing organism. Thus, bio-fertilizers are cultures of appropriate species of microbes that have the capability of fixing atmospheric nitrogen such as Rhizobium species in leguminous crops and Azotobacter and Azospirillum in non-leguminous crops. The phosphate solubilizing bacteria (PSB) and phosphate mobilizing fungi (VAM) have been found very effective in making 23 unavailable soil phosphorus available to plants. The legume – Rhizobium association can fix nitrogen in the range of 40-120 kg/ha under optimum conditions • Symbiotic N-fixing bacteria-Rhizobium • Asymbiotic N- fixing bacteria-Azotobacter, Azospirillium • P-solubilizing fungi-mycorrhizae-VescicularArbuscularMycorrhizae (VAM) • P-solubilizing bacteria • Anabaena-Azolla association Biofertilizers viz: Rhizobium, Azotobacter, Azospirillum, PSBAzolla.

Agro industrial wastes: List of industries with potentiality to produce large quantities of organic wastes. • Sugar industry • Rice industry • Cotton industry • Tea/tobacco industries • Silk industry • Paper industry • Fruit processing industry • Coconut based industries • Areca based industries • Coffee pulping industry • Jute industry • Vegetable market waste compost.

Bio conversion of under-utilized sources:

(a) By-products of Slaughter Houses

(i) Bone meal: An estimated quantity of 4.5 lakh tons of animal bones are available in our country, annually out of which nearly 1.4 lakh tons are collected and crushed. Remaining quantities of bones are yet to be utilized.

(ii) Meat meal: Slaughter houses produce large quantity of meat not normally used as food. It is estimated that nearly 1.2 lakh tons of decomposable animal wastes are generated from slaughter houses.

(iii) Leather wastes: Although major part of arable leather reaches leather industry for processing, sizeable leather wastes are generated from slaughter houses estimating to about 5000 tons annually.

(b) Bio-Gas Slurry

An estimated number of 20 lakh bio-gas plants are working in India. Bio-gas plants being a digester to convert crop residues and animal dung into combustible gas generates large quantities of bio slurry. It is rich in plant nutrients. On an average it contains 1.4 per cent N, 0.8 per cent P2 O5 and 0.6 per cent K_2O.

(C) Poultry Manure

Each poultry bird on an average produce 0.025 kg litter (excreta) per day on dry weight basis, 2.8 million tons of poultry manure is produced annually in our country. The average manorial value of this poultry manure works out to 27,553 tons of N, 36,772 tons of P_2O_5 and 21,941 tons of K_2O. Use of poultry manure will open newer options in the field of agriculture.

Minerals

The mineral based materials like rock phosphate, gypsum, lime stone, calcium chloride, Sodium chloride Magnesium rock and chalk etc can be applied in limited quantities when there is absolute necessity. The product such as Saw dust, wood shaving from untreated wood and plant preparation and extracts products are permitted for use in manuring/soil conditioning in organic field.

Cultural Methods

Organic sources of nutrients augmented with different cultural methods are more effective. There is no involvement of additional materials but a mere shift in method of cultivation or pattern or system of cropping. These methods have been part of our agriculture since a long time and some have been introduced lately. These methods contribute to increasing nutrient concentration. Important cultural methods include:

• Effective and useful rotation of crops • Ideal intercropping with complimentary interactions • Alley cropping with N-fixing trees • Practice of homestead farming • Soil and water conservation practices • Fallowing • Establishment of vegetative bunds.

Weed Management

The effect of weedicides on soil health has been totally ignored which is causing serious changes in soil ecology. Application of weedicides is not permitted in organic farming. Weed management is becoming ineffective due to emerging herbicide resistance in weeds. Scientists have come out with the technology, which reduces the weed density and increases the crop competition with weeds. Control of weeds including *Phalaris minor* can be done to a larger extent through changing the crop dynamics and timely sowing. Mechanical weeding is highly beneficial for crop establishment. Cultivation practices may be devised in such a way that help the crops in capturing the resources easily and grow vigorously in comparison to weeds. Biological control of weeds can be highly effective. However, under Indian conditions; the approach has to be evolved. Use of locally available mulching materials or polythene sheets to reduce weed growth moisture conservation can be effectively employed in vegetable cultivation Novel preparations such as biodynamic formulations etc can be used in appropriate quantity. Use of Biodynamic preparations, such as BD-500 and BD-501 as foliar spray has also been found to be effective in growth promotion.

References

Barik, A., Sarkar, N. (2017, November 8-11). Organic Farming in India: Present Status, Challenges and Technological Break Through. In: 3rd International Conference on Bio-resource and Stress Management, Jaipur, India.

Bhardwaj, M., Dhiman, M. (2019). Growth and performance of organic farming in India: what could be the future prospects? Journal of Current Science, 20: 1–8.

Chandrashekar, H.M. (2010). Changing Scenario of organic farming in India: an overview. International NGO Journal, 5: 34–39

Chopra, A., Rao, N.C., Gupta, N., Vashisth, S. (2013).Come sunshine or rain; organic foods always on tract: a futuristic perspective. International Journal of Nutrition, Pharmacology Neurological Diseases, 3: 202–205.

Deshmukh, M.S., Babar, N. (2015). Present status and prospects of organic farming in India. European Academic Research, 3: 4271–4287

Reddy S.B. (2010). Organic farming: status, issues and prospects—a review. Agricultural Economics Research Review, 23: 343–358

14

Insect Pests of Rose & Their Management by Integrated Pest Management

Saurav Gupta, Vinod Gupta, Neerja Sharma and A.K. Sinha

KVK, Samba, SKUAST-Jammu, Jammu and Kashmir, India

Pest

A pest is any animal or plant detrimental to crops, livestock and forestry. The term is also used of organisms that cause a nuisance, such as in the home. An older usage is of a deadly epidemic disease, specifically plague. In its broadest sense, a pest is a competitor of humanity, means disease causing microbes, plant devouring insects, macro-fauna like rats and moles, weeds, off type plants and even human beings are pests in perspective of plants.

IPM

IPM is a pest management system that in the context of associated environment and population dynamics, utilizes all the appropriate techniques to minimize the pest population levels below those causing economic injury. Integrated Pest Management (IPM) includes cultural Practices, Host Plant Residence, Mechanical control, Biological control, Use of Bio-pesticides, botanicals and Chemical control.

1. Cultural Methods

Cultural methods includes Sanitation, Tillage and inter-cultivation, Time ofsowing, Plant population, Manuresand fertilizers, Water management, Habitat diversification Agronomic adjustments, necessary for higher yield, are at the same time are directed at prevention, mass multiplication and spread of pests by modifying the crop microclimate.

Physical methods include Treatment of Planting Material

Firstly, the first priority should be healthy planting material. If we are not 100% sure about its being healthy then it should be treated properly to prevent further build up of diseases. E.g. Hot water treatment Corms and cormels of Gladiolus, Bulbs of Tuberose etc.

a. Irrigation: Irrigation of the field prevents the termite attack.

b. Drainage: Proper drainage prevents the build up of rot fungus and bacterium and also several insect pests. E.g. grey moulds of all ornamental plants can be prevented by sanitation and drainage, Botrytis bud and twig blight of rose

c. Sanitation: It includes removal or destruction of breeding refuges and over wintering of pests. Seed material, farm yard manure etc. carrying insect eggs or its stages of development should be carefullyscreened before their use. Destruction of alternate hosts minimizes pest population build up. E.g. Rose black spot/Die back/leaf spot/rust, basal rot of Carnation and downy mildews and all wilts and rots.

d. Tillage and inter-cultivation: Ploughing and inter-cultivation brings about unfavourable conditions for multiplication of pests as well as diseases and weeds. E.g. Proper drainage in the field prevents the build-up of wilts and rots. Quiescent stages (pupae) of harmful organisms will be exposed to dehydration or to predation by birds and other stages may be mechanically damaged or buried deep in the soil. E.g. Marigold inter-cultivation prevents nematode growth.

e. Cultivar selection: Cultivars with high yield potential and quality without resistance to pests and diseases are the main causes of frequent epidemics and mass multiplication of pests and diseases. A large number of cultivars resistance/tolerance to pest and diseases has been developed to suit different agro- ecosystems. Selection of such cultivars can bring down the losses considerably.

f. Time of sowing: As weather influences developmental rhythm of plants as well as growth and survival of pests and diseases, serious setback occurs when the weather conditions are such as to bring about coincidence of susceptible growth stages with highest incidence of pests and diseases. Therefore, adjustment in sowing dates is often resorted to as an agronomic strategy to minimize the crop losses.

g. Habitat diversification:Many pests prefer feeding on a particular plant or others. This preference may be exploited to reduce the pest load on crop. Crop rotation, intercropping, traps cropping and strip cropping can bring down the pest load considerably

h. Trap cropping: Trap cropping is the planting of a trap crop to protect the main cash crop from a certain pest or several pests. The trap crop can be from the same or different family group, than that of the main crop, as long as it is more attractive to the pest. Trap cropping Perimeter/ Border trap cropping Row intercropping.

Advantages of trap cropping: Lessens the use of pesticide. Lowers the pesticide cost Preserves the indigenous natural enemies. Improves the crop's quality helps conserve the soil and the environment.

2. Mechanical Control

a. Hand picking method:Hand picking of eggs and larvae of insects is very effective method. This method of pest control is useful if the crop is in a small area. Pour a small amount of kerosene in a polythene bag and pick up the larvae during evening hours and put it in the bag. The pests can be controlled this way without the use of any chemicals. This should be done when the pest numbers are low. Hand picking of larva: Wild grasses and weeds should be removed from the field bunds and field, since, these are the favourite egg laying spots of the pests.

b. Bird Perches: Install 'T' shaped bird perches which are long dried twigs @15-20 per acre. These attract birds for resting and the resting birds devour the larvae in the field. Bird perches are resting places for predatory birds to rest and to look for preys; such as insect pests of cotton, peanuts, and cowpeas. Predatory birds prefer to look for prey in field crops where they have places to rest. To make bird perches, use bamboo or wooden poles or tree branches. Erect either of these at regular intervals in the field. To have live bird perches within the field, plant Setaria species (foxtail cultivars). These plants are found to be attractive to predatory birds. The birds feed on their seeds. In cotton field, plant Setaria in every 9th or 10th row of cotton. Once the birds are on the field, they prey on cotton bollworms and other insects.

c. Installation of Yellow/Blue/White Sticky traps: Installation of Yellow/Blue/White sticky traps @50/ha are very effective for monitoring/controlling small sucking pests like Aphids, Thrips, Whiteflies, bugs etc. Castor oil smeared yellow colour empty tins or plates are kept in the field. White flies get trapped on these sticky traps. These are wiped out every day and castor oil is applied again. Bright blue traps are for monitoring thrips, Bright white traps are for monitoring flea beetles and tarnished plant bugs.

d. Pheromones Traps: Pheromones are ectohormons secreted by an organism, which elicit behaviouralresponsesfrom other members/sex of its own species. These are extremely selective, nontoxic, highly biodegradable and effective at low application rates. Synthetic sex pheromones are commercially available and are used for surveillance,monitoring and control of many Lepidopterous pests such as spotted bollworm, tobacco caterpillar, potatotuber moth, diamond back moth and leaf folderetc.

e. Fairomones: These are volatile compounds that evoke behavioural response adaptively favourable to the receiver. Fairomones are released either by the host plant or by the host insects. While former issued by the pest and natural enemies to locate their habitats, the later is used for prey finding and parasitisation/preying. Fairomones from host plant can be effectively used to mass trap pest species as well as for monitoring. Use of fairomonal compounds to increase the efficiency of the predator C. carnea and the egg parasitoid T chilonis had been successfully demonstrated.

f. Light trap: Light trap is a device used at night in the field to collect moths and other flying insects such as: armyworm Bugs, Cutworm Flies, Leafhoppers, Planthoppers, Stem borers etc.

3. Biological Control

Biological control basically means, "The utilization of any living organism for the control of insect-pest, diseases and weeds". This means use of any biotic agents for minimizing the pest population either directly or indirectly. Classical biological control involves deliberate and natural establishment of natural enemies in areas where they did not previously occur. In addition to deliberate introduction of biocontrol agents, proper attention needs to be given for conservation and augmentation of natural enemies that already exist in an area. This should be treated as important as many insect predators are much more susceptible to insecticides than the pests they attack. Biological control agents for insect pests are available in nature abundantly and work against crop pests naturally called as Natural Control. E.g.Powdery mildew of rose -The fungal antagonist Sporothrixflocculosa has been reported to be highly effective in controlling the disease (Belanger et al., 1994). Tilletiopsispallescens, a naturally occurring ballistospore forming yeast isolated from mildew infected leaves has bean proved to be effective against the fungus. The bio agent reduces the incidence of disease by 97-98% (Na et al., 1997)

Bio Intensive Pest Management (BIPM) The pest management programme where natural enemies of the crop pests form the core component is designated as (BIPM). The most commonly used bio-agents in BIPM are broadly classified into three categories:

1. Parasitoid 2. Predators 3.Microbes.

a. Parasitoid

These are the insects, either equal or lesser than the size of the host insect (pest) and always require passing at least one stage of their life cycle inside the host system. Due to their high multiplication rates they are of vital importance in the biological control of insect-pests. Trichogramma Egg parasite against eggs of gram pod borer, rice stem borer, castor semilooper, cabbage diamond back moth etc.Braconhebetor against insect-pest of coconut and sugarcane Brachymeria against the pupae of several pests of plantation crops.

b. Predators

These are those insects, which are generally bigger than the host and feed on several of the pests by predating upon them externally. They will be consuming several of the insect- pests during their life cycle and hold a key role in minimizing pest population in field conditions. Predator Prey Chilizonus, Cryptolaemus, Scymnus, Meoichlus and Pharoscymnus. mealy bugs, coccids, scales and mites on citrus, grapevine and guava Mirid predator Crytorhinus brown plant hoppers Insect predator Chrysopa aphids, mealy bugs and young caterpillars. Chrysoperlasp several of the soft bodied insects such as aphids, leaf hoppers etc; ladybird beetle aphids and mealy bugs; spiders varied number and types of insects especially in ecosystem.

c. Microbial organisms

These are those micro- organims which are capable of causing diseases in insects as a result they lose their appetite, subjected to several physiological disturbances, leading to the ultimate death ofthe insect. Nematode pest management by using biotic agents is also one of the most promising areas and gaining much desired importance inthe current scenario of organic farming. a) Insect viruses b) bacteria c) Entomo Pathogenic Fungi d) Entomo Pathogenic Nematodes and other organisms like protozoans and Rickettesia etc. several antagonistic fungi and bacteria being successfully used in minimizing the plant disease incidence.

i. **Viruses:** Nuclear polyhedroses viruses (NPV): They cause typical "tree top disease" in insects. The infected insect loose appetite becomes restless and reaches the apical portion of the plant due to oxygen depletion inside the insect system. Later the infected insect will be dying by hanging itself in an inverted 'V' shape on the apical portion of the plant. Effective against lepidopteran insects in different crops. HaNPV is used for the management of *Helicoverpaarmigera* while SlNPV is meant for Spodopteralitura.. Entomopathogenic viruses highly specific to host insects which make them exceptionally safe to non-target organisms and nature.

ii. **Bacteria and Fungi:** Bacillus thuringensis: This bacterium is highly effective against several insect-pests belonging to order Lepidoptera. They cause disease due to which insect turns black and die. MetarhiziumThese are used against gram pod borer, tobacco anisopliae, caterpillar and sucking pests like thrips, aphids and Beauveria mealy bugs. The fungi develop hyphae inside insect bassiana and system resulting in the death of the insect due to Verticillium mechanical congestion. The mode of action makes lecanii these organisms to perfectly suit to the needs of organic farming. In certain cases they produce toxins to kill the insect.

iii. **Nematodes and Protozoa:***Heterorhabditis* sp and *Steimernema* sp: These nematodes harbor certain bacteria which act as toxins to insect systems. *Variomorpha*sp: These were found effective against insect pests and can effectively incorporated as tools in organic farming.

4. Bio-pesticides

Natural occurrence of diseases caused by micro- organisms is common in both insects and weeds and is a major natural mortality factor in most situations. Use of micro-organisms for pest control involves their culture in artificial media and later introduction of larger amounts of inoculums in to the field at appropriate time. Many fungi and bacteria can be handled in this way but insect viruses have the limitation that they have to be raised in living insects. As the biocontrol agents (microbial pathogens) are applied on the targeted pests in much the same way as chemical pesticides, they are often termed as bio-pesticides or natural pesticides.

a. Bacillus thuringensis, a bacterial pathogen infesting a wide range of insect pests, is the most common microbial insecticide in use today. It is used against caterpillars that attack a wide range of crop. Unlike most other chemical insecticides, it can be used on edible products up to the time of harvest. It is selective in action and does not harm parasites, predators or pests. The bacteria come in several commercial formulations such as Dipel, Delfin, Halt, Spicturin, Biolep, BioAsp etc. Another bacterium B popilaleis also commonly available against white grub Popillae japonica and Hototricha sp. Amongst insect pathogenic fungi, commercial preparations of Verticilliumlecanii are available for the control of aphids, thrips and white fly under glass house conditions.

5. Botanicals

Some weeds like Lantana, notchi, tulsi, adathodaetc act as natural repellant to many pests. Trees like pungam, wood apple, anona and their byproducts have excellent insecticidal value in controlling diamond back moth, heliothis, white flies, leafhoppers and aphid infestation. Most commonly used botanicals are neem (Azadirachtaindica), pungamia (Pungamiaglabra) and mahua (Madhucaindica).

Neem seed kernel extract (2 to 5%) has been found effective against several pests including rice cutworm, diamond back moth, rice BPH, rice GLH, tobacco caterpillar, aphids and mites.E,g

a. The pesticidal ingredients of neem formulations belonging to general class of natural products called triterpenes, more specifically, limonides. They act as repellents and also disrupt growth and reproduction in insects. Commonly known limonoidsareazadirachtin, meliantriol, salannin, nimbin and nimbidin. The efficiency of vegetable oils in preventing infestation of stored product pests such as bruchids, rice and maize weevils has been well documented. Root extracts of asparagus work as a nematicide for plant parasitic nematodes. Similarly leaf extracts of many plants can inhibit a number of fungal pathogens.

b. Neem as pest repellent: Take neem leaves or Neem cake or Neemkernelsand ground it well and place it in a pot. Add twice the volume of water and tie the mouth of the pot with a cloth. Leave it as such for three days. Then, place the pots on all the 4 corners of a field. In the evening, open the mouth of the pots. The foul smell emanating from the neem products prevents entry of pests into the field.

A. Diseases of rose and management

1. **Die back (*Diplodiarosarum*):**

Symptoms: Blackening and drying of shoots from tip downward.

Control

a. Resistant varieties are Bhim, Blue Moon, etc.

b. Pruning of dried shoots and pasting of blitox or Bordeaux paste on the cut portions.

c. Spraying of blitox @0.2%.

2. Black spot (*Diplocarponrosae*)

Symptoms: Conspicuous circular black spots with fringed margins, etiolation of leaves and defoliation.

Control

a. Phytosanitary measures like removal of fallen leaves

b. Spraying of ferbam @0.2%, or captan @0.2%, or bavistin @0.1%.

3. Powdery mildew (*Sphaerothecapannosa*):

Symptoms: White powdery patches on all aerial parts.

Control

a) Planting of resistant varieties Ranjana, Kalawati, Dr. HomiBhava, etc.

b) Spraying of karathane @0.05% or topsin-M @ 0.1%, or bavistin @0.1%, or sulphur 80WP @0.1%.

4. Rust (*Phragmidium* sp.):

Symptoms: Black pustules, rusty appearance on leaves.

Control

a. Spraying of vitavax @0.1% during March to April.

b. Use of healthy plantlets.

5. Bud and twig blight (*Botrytis cinerarea*):

Symptoms: Brownish patches on petals, rotting.

Control

a) Spraying of bavistin @0.1%

b) RaktaGandha, Purna, Elizabeth are resistant.

6. Downy mildew (*Peronosporasparsa*):

Symptoms: Cottony growth on lower surface of leaves, light yellow irregular patches on surface.

Control

a. Sanitation in the field

b. Spraying of ridomil @0.1%

c. Spraying of blitox @0.2%.

7. Rose mosaic disease, Prunus necrotic ring spot:

Symptoms: Mosaics of alternate yellow and green areas on leaves.

Control

a) Meristem tip culture – for getting disease free plants.

b) Chemotherapy using dye stuff/virazole/charcoal.

c) Control of vectors by insecticides.

B. Insect pests of rose, their damages and management:

1. Red scale (*Aonidiellaaurantii*)

Damages: Reddish brown encrustations on shoots; both young and adult scales suck the sap from shoots.

Control

a. Pruning of infested plant parts.

b. Planting scale free rose plants, iii) Spraying of dimethoate @0.05% or malathion @ 0.25%.

Scales on Rose stems

2. Aphid (*Macrosiphumrosae*)

Damages: Aphids cluster around growing shootsand buds, and suck cell sap. Affected plant parts get deformed.

Control

a) Spraying of neem or pongamia oil @ 2%,

b) Spraying of malathion @0.15.

3. Termite (*Microtermesobesi*)

Damages: Damage root systems and the plant die.

Control

a) Deep ploughing of the field,

b) Drenching soil with Chlorpyriphos @ 0.05%

c) Irrigation on time.

4. Hairy caterpillar (*Euproctisfracterna*):

Damages: Larvae feed on the leaves, buds, and flowers resulting in defoliation and flower loss.

Control

a) Collection and destruction of egg masses on lower side of leaves,

b) Light traps attract adult moths, iii) Spraying quinalphos @0.05%.

5. Thrips (*Rhipiphorothripscruentatus&Scirtothripsdorsalis*):

Damages: Thrips damage new flush by sucking sap from tender parts; Scirtothrips is a problem in polyhouse roses.

a) Control:

a) Spraying of neem or pongamia oil @ 2%,

b) Spraying of imidacloprid @0.4 ml/litre.

6. Chaffer beetle (*Adoretus* spp.)

Damages: Beetles emerge from the soil after dusk during monsoon and feed on leaves, buds and tender shoots. Grubs feed on the roots.

Control

a) Deep pluoghing to kill the grubs and pupae,

b) Spraying of neem seed kernel extract @2.5%,

c) Spraying of chlorpyriphos @0.05% during evening,

d) Application of Cartap in soil @ 22.0 kg /ha to control grubs.

7. Two spotted spider mite (*Tetranychusurticae*):

Damages: Discolouration, bronzing and drying of leaves.

Control

a) Severely infested plant parts should be pruned.

b) Spraying of mitac @0.05% or cascade @0.15%.

C. Nematode pests in rose and their management

1. Lesion nematode (*Pratylenchusconvalleriae*)

Symptoms: Chlorosis of leaves, less flowering, destruction of feeder roots.

Control

a) Soil disinfection with methyl bromide/parathion,

b) Growing Tagetes and Gaillardia as trap crops,

c) Rose variety Major is resistant to the nematode.

2. Meloidogyne sp.

Symptoms: Stunted plant growth and galling on roots.

Control

a) Flooding with 0.1% nemaphos.

b) Intercropping with marigold.

c) Application of carbofuran @1 g a.i. per m^2.

d) Application of neem cake @ 1 ton/ha.

References

Ananthakrishan, T. N. Thrips (Thysanoptera) in agriculture, horticulture and forestrydiagnosis, bionomics and control. J. Sci. Ind. Res., 1971, 30, 113–146.

Boll, R., Marchal, C., Poncet, C., & Lapchin, L. Rapid visual estimates of thrips(Thysanoptera: Thripidae) densities on cucumber and rose crops. J. Econ. Entomol., 2007, 100(1), 225–232.

Gahukar, R. T. New record of thrips attacking roses in central India. J. Insect Sci., 1999, 12(1), 89.

Mehrparvar, M., Mobli, M., &Hatami, B. Seasonal population fluctuations of the rose aphid, *Macrosiphumrosae* (L.) (Hemiptera: Aphididae), on different cultivars of roses and Nastaran. J. of Sci. and Tech. of Agric. Res. and Natural Resources, 2008, 12(45B), 711–719.

Rajkumar, M., Reddy, K. L., &Gour, T. B. Thrips and mites infesting roses. Insect Environ., 2004, 10(1), 27–28

15

Integrated Scheme for Agricultural Marketing A Review

Sheema Khan and Poonam Parihar

Division of Agricultural Extension Education, Faculty of Agriculture SKUAST-Jammu, Jammu and Kashmir, India

Abstract

In the present era, the farmers are not only achieving self sufficiency in production but are also producing surplus. Farmers have been able to take advantage of many schemes launched by the governments from time to time in order to achieve this goal. However, in the present context, the main challenge that concerns a farmer is the marketing of his produce in order to achieve maximum profit. Realizing this need, many market-oriented schemes are being launched by the state and central governments. It is well known that there are several challenges in the marketing sphere in general such as lack of adequate and well-equipped warehouses, lack of grading and standardization of the produce, inadequate transports facilities, presence of a large number of middlemen, malpractices in unregulated markets, inadequate market information and credit facilities available with farmers, etc. (Kumar, 2018). These challenges have brought to the fore, the need for providing farmers with access to competitive markets with adequate infrastructure including cold chain logistics, to enable them to realize better prices on the one hand and providing nutritious food to consumers at stable and affordable prices on the other. With this objective in view, the Government of India on 13th November, 2013 approved the proposal of Department of Agriculture & Cooperation for continuation and integration of on-going Central Sector Schemes as Integrated Scheme for Agricultural Marketing (ISAM) during the XII Plan.This scheme comprises of certain sub-schemes: (a) Agricultural Marketing Infrastructure (AMI), (b) Marketing Research and Information Network (MRIN), (c) Strengthening of Agmark Grading Facilities (SAGF),

(d) Agribusiness Development (ABD) through Venture Capital Assistance (VCA) and Project Development Facility (PDF), (e) Choudhary Charan Singh National Institute of Agriculture Marketing (NIAM).

Keywords: Agricultural marketing, Integrated scheme for agricultural marketing, Agmarknet, Agmark, e-NAM, Agricultural marketing infrastructure.

Introduction

The marketing systems and post-harvest marketing infrastructure have not been able to keep pace with the growing production and marketable surplus. This has brought to the fore, the need for providing farmers with access to competitive markets with adequate infrastructure including cold chain logistics, to enable them to realise better prices on the one hand and providing nutritious food to consumers at stable and affordable prices on the other. With this objective in view, the Government of India on 13th November, 2013 approved the proposal of Department of Agriculture & Cooperation for continuation and integration of on-going Central Sector Schemes as Integrated Scheme for Agricultural Marketing (ISAM) during the XII Plan (2012-2017). The ISAM have the following five sub schemes: (i) Agricultural Marketing Infrastructure (AMI) [the existing schemes of Grameen Bhandaran Yojana (GBY) and Development/Strengthening of Agricultural Marketing Infrastructure, Grading and Standardization (AMIGS) will be merged as AMI] (ii) Marketing Research and Information Network (MRIN) (iii) Strengthening of Agmark Grading Facilities (SAGF), (iv) Agribusiness Development (ABD) through Venture Capital Assistance (VCA) and Project Development Facility (PDF) and (v) Choudhary Charan Singh National Institute of Agriculture Marketing (NIAM) (www.agricoop.nic.in).

Market information is needed by farmers in planning production and marketing, and equally needed by other market participants in arriving at optimal trading decisions. The existence and dissemination of complete and accurate marketing information is the key to achieve both operational and pricing efficiency in the marketing system.Advancement in Information and Communication Technology (ICT) has made the world a smaller place and a larger market at one go. To fully utilize the new emerging marketing opportunities for the benefit of farming community,establishment of an ICT based "Agricultural Marketing Information Network" in the country has become inevitable.

Ministry of Agriculture has launched the ICT based Central Sector Scheme of Agricultural Marketing Information Network (AGMARKNET), in March, 2000, to link important agricultural produce markets spread all over the country and the State Agriculture Marketing Boards and Directorates. The project is being executed with the technical support of National Informatics Centre (NIC). In order to strengthen interface with farmers and other beneficiaries, AGMARKNET portal (http://agmarknet.nic.in) has been evolved. The AGMARKNET portal serves as a single window for accessing websites of various organizations concerned with agricultural marketing (https://www.nic.in/).

Warehousing plays a vital role in promoting agricultural marketing, rural banking and financing and ensuring food security in the country. A network of scientifically constructed storage infrastructure will help farmers avoid distress sale of their produce by availing marketing credit through pledge financing and also reduce quantitative and qualitative post-harvest losses.

SFAC venture capital scheme assists agripreneurs to make investments in setting up agribusiness projects through financial participation, and provide financial support for preparation of bankable Detailed Project Reports (DPRs) through Project Development Facility (PDF) (http://sfacindia.com/).

SAGF sub scheme of the ISAM Scheme is an ongoing plan scheme to support grading and marking of agricultural produce, which involves framing of grade standards and certification of agricultural commodities included in the Schedule of the Agricultural Produce (Grading and Marking) Act, 1937.

The Government of India launched an electronic-National Agriculture Market (e-NAM): a pan-India online trading portal on 14 April 2016, to improve price discovery in real-time by eliminating the information asymmetry between buyers and sellers. It connects surplus production regions with deficit regions through an online platform, which may lead to better market competition, and thus better prices for farmers for their produce. It also provides an option for buyers to access products across markets (Venkatesh*et al.,* 2021).

Discussion

Marketing

According to the National Commission on Agriculture, agricultural marketing is a process which starts with a decision to produce a saleable farm commodity, and it involves all the aspects of market structure or system, both functional and institutional, based on technical and economic considerations, and includes pre- and post-harvest operations, assembling, grading, storage, transportation and distribution.Agricultural marketing covers the services involved in moving an agricultural product from the farm to the consumer. Numerous interconnected activities are involved in doing this, such as planning production, growing and harvesting, grading, packing, transport, storage, agro- and food processing, distribution, advertising and sale (Jabeen and Jabeen, 2021).

1.1 Challenges of Present Marketing System

1. Improper warehouses
2. Lack of grading and standardization
3. Inadequate transport facilities
4. Presence of a large number of middlemen
5. Malpractices in unregulated markets
6. Inadequate market information
7. Inadequate credit facilities

Integrated Scheme for Agricultural Marketing

The above challenges have brought to the fore, the need for providing farmers with access to competitive markets with adequate infrastructure including cold chain logistics, to enable them to realize better prices on the one hand and providing nutritious food to consumers at stable and affordable prices on the other. With this objective in view, the Government of India on 13th November, 2013 approved the proposal of Department of Agriculture & Cooperation for continuation and integration of on-going Central Sector Schemes as Integrated Scheme for Agricultural Marketing (ISAM) during the XII Plan.

Objectives

i. To promote creation of agricultural marketing infrastructure by providing backend subsidy support to state, cooperative and private sector investments.

ii. To promote creation of scientific storage capacity and to promote pledge financing to increase farmers' income.

iii. To promote Integrated Value Chains (confined up to the stage of primary processing only) to provide vertical integration of farmers with primary processors.

iv. To use ICT as a vehicle of extension to sensitize and orient farmers to respond to new challenges in agricultural marketing.

v. To establish a nation-wide information network system for speedy collection and dissemination of market information and data on arrivals and prices for its efficient and timely utilization by farmers and other stake holders.

vi. To support framing of grade standards and quality certification

Sub Schemes of ISAM

a) Agricultural Marketing Infrastructure (AMI)

b) Marketing Research and Information Network (MRIN)

c) Strengthening of Agmark Grading Facilities (SAGF)

d) Agribusiness Development (ABD) through Venture Capital Assistance (VCA) and Project Development Facility (PDF)

e) Choudhary Charan Singh National Institute of Agriculture Marketing (NIAM)

Agricultural Marketing Infrastructure (AMI)

- To develop agricultural marketing infrastructure for effectively managing marketable surplus of agriculture including horticulture and of allied sectors including dairy, poultry, fishery, livestock and minor forest produce.
- To promote innovative and latest technologies in agricultural marketing infrastructure.
- To promote competitive alternative agricultural marketing infrastructure by encouraging private and cooperative sector investments.
- To promote direct marketing so as to increase market efficiency through reduction in intermediaries and handling channels thus enhancing farmers' income.
- To promote creation of scientific storage capacity for storing farm produce, processed farm produce and agricultural inputs etc. to reduce post-harvest and handling losses.

Eligible Beneficiaries

Assistance under the sub scheme will be available to:

1. Individuals, Group of farmers/growers, Registered Farmer Producer Organisations (FPOs).
2. Partnership/Proprietary firms, Companies, Corporations.
3. Non-Government Organizations (NGOs), Self Help Groups (SHGs).
4. Cooperatives, Cooperative Marketing Federations.
5. Autonomous Bodies of the Government, Local Bodies (excluding Municipal Corporations for storage infrastructure projects), Panchayats.
6. State agencies including State Government Departments and autonomous organization/State owned corporations such as Agricultural Produce Market Committees & Marketing Boards, State Warehousing Corporations, State Civil Supplies Corporations etc.

Channelizing Agencies for Release of Subsidy

Subsidy will be released through NABARD for the projects financed by Commercial, Cooperative, Regional Rural Banks, Agricultural Development Finance Companies (ADFCs), State Cooperative Banks (SCBs), State Cooperative Agricultural and Rural Development Banks (SCARDBs), Scheduled Urban Cooperative Banks, Scheduled Primary Cooperative Banks (PCBs), North Eastern Development Financial Corporation (NEDFi), other institutions eligible for refinance by National Bank for Agriculture and Rural Development (NABARD) or any other Financial Institution such as State Financial Corporations (SFCs) approved by DAC.

Subsidy will be released through NCDC for projects financed by NCDC or by cooperative banks recognized by NCDC in accordance with NCDC's eligibility guidelines.

Subsidy will be released directly by DAC in case of projects promoted by State agencies whichpropose to invest their own funds without availing any loan.

In addition to NABARD and NCDC, DAC may authorize Small Farmers Agribusiness Consortium (SFAC) for release of subsidy for credit linked projects of private sector/ State agencies including FPOs.

Marketing Research and Information Network (MRIN)

Market information is needed by farmers in planning, production and marketing of their produce and equally needed by other market participants in arriving at optimal marketing decisions. The availability and dissemination of complete and accurate marketing information is key to achieve both operational and pricing efficiency in the marketing system.

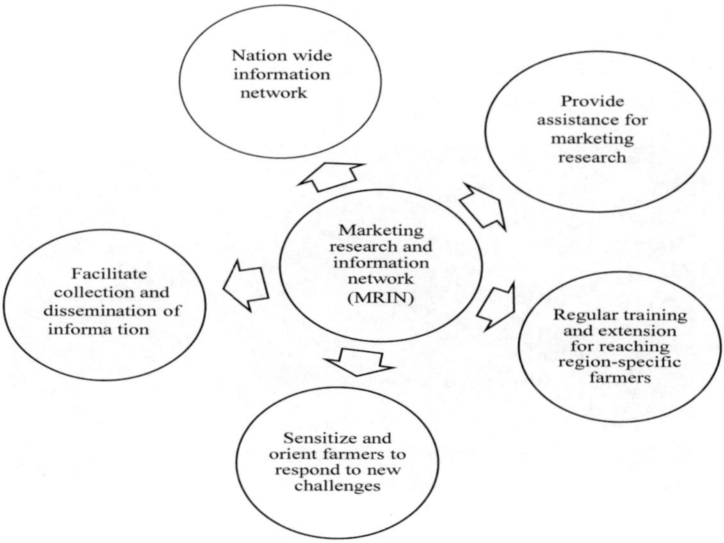

In order to strengthen interface with farmers and other beneficiaries, AGMARKNET portal (http://agmarknet.nic.in) has been developed. More than 2200 markets are regularly reporting price and arrivals related data which is being disseminated through the portal. The AGMARKNET portal also serves as a single window for accessing websites of various other organizations concerned with agricultural marketing. It provides weekly arrivals and price trend analysis for important markets in respect of major agricultural commodities transacted.

The State Agricultural Marketing Boards/Directorates/NIC will be provided funds under the sub scheme from Grant-in Aid.

1. Supply of Computers and other peripherals (Annexure-II) as well as installation will be implemented through the vendors empaneled by IT Division of DAC, or/and other sources such as NIC/APMC, State Departments/Agencies in concerned market nodes as per the pricing norms stipulated by the IT Division of DAC / NIC or at DGS & D rates.

2. Supply/procurement of electronic instruments such as Mobile handset of particular specification and required software to all selected markets will be carried out as per the decision of the sanctioning committee and in conformity with GFR norms.

3. Training of market personnel in handling of hardware and software and refresher training will be undertaken by NIC/any other service providers.

The cost towards training will be reimbursed to NIC / any other service providers. Supplementary training needs for technical aspects of data collection will be addressed by NIAM.

4. Nodes reporting data for more than 20 days in a month will be ascertained from the website's Performance Monitoring System and on certification by NIC. Incentives would be released on Quarterly basis.

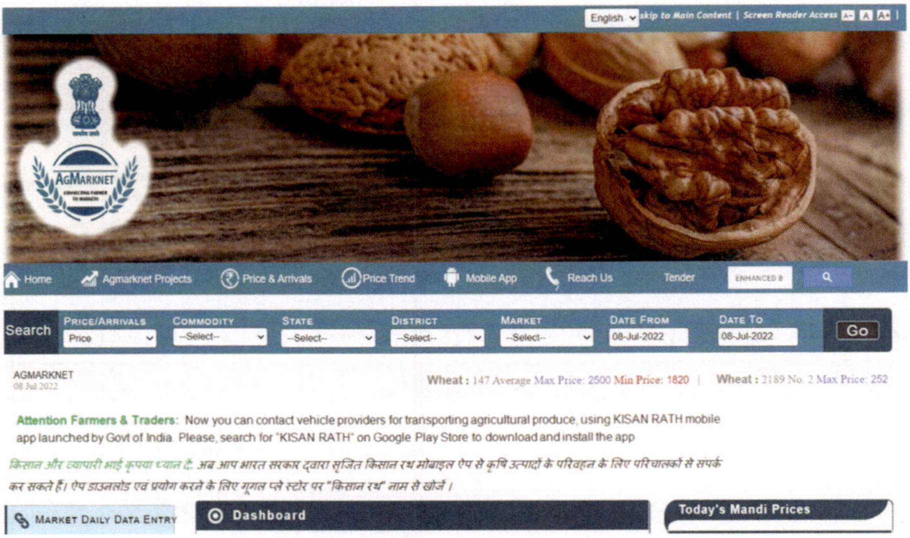

Strengthening of Agmark Grading Facilities

1. To help farmers get better and remunerative prices by grading of their produce.
2. To frame grade standards of agricultural commodities as per the provisions in Agricultural Produce (Grading and Marking) Act,1937.
3. To implement AGMARK certification programme for commodities for which grade standards are notified for domestic trade and for exports.
4. To analyze research samples for creating analytical data base for the framing/revision of grade standards of agricultural commodities.

Implementing Agency

Directorate of Marketing & Inspection (DMI) in the Department of Agriculture & Cooperation,Ministry of Agriculture, is the Nodal authority for certification.

Infrastructure for the Grading & Standardization Programme

DMI has infrastructure of 11 Regional Offices and 26 Sub-offices spread all over the countryto implement the certification programme. Head Office at Faridabad attends to the policy matters and framing/revision of standards based on the Reports of research projects carried out at Central Agmark Laboratory(CAL). Eleven Regional Agmark Laboratories (RALs) spread all over the country provide analytical support for the analysis of research samples and check samples drawn from the authorized packers' premises and market. Central Agmark Laboratory at Nagpur is the apex laboratory. The RALs at Mumbai & Chennai and CAL at Nagpur are accredited with National Accreditation Board for Testing & Calibration Laboratories (Department of Science & Technology).

Table 1. AGMARK Online Certification

S.No.	Category	Number
1.	Total online users	12922
2.	Total applications received	6287
3.	Total applications granted	3102
4.	Certificate of authorization	4604
5.	Chemists in 12 labs	90
6.	Approval of laboratory	298
7.	Total revenue	24655 K

Certification for Domestic Market

The approved Chemist of the CA holder analyses the processed commodity for determining the grade standard and gets it packed in his presence.Check samples are drawn from premises of the authorized packers and market and these are analyzed in RALs to keep check on the certification programme. Action to warn and suspend/cancel the Certificate of Authorization is taken, if any, check sample is found not conforming to prescribed standards.

Certification for Exports

DMI is attending to certification of fruits and vegetables for exports through approved laboratories. 13 laboratories spread all over the country are approved to attend to grading and certification.

Agribusiness Development (Abd) Through Venture Capital Assistance (Vca) and Project Development Facility (PDF)

a. To facilitate setting up of agribusiness ventures in close association with all Financial Institutions notified by the Reserve Bank of India where the ownership of the Central/State Government is more than 50% such as Nationalized banks, SBI & its subsidiaries, IDBI, SIDBI, NABARD, NCDC, RRBs & State Financial Corporations.

b. To catalyze private investment in setting up of agribusiness projects and thereby providing assured market to producers for increasing rural income & employment.

c. To assist farmers, producer groups, and agriculture graduates to enhance their participation in value chain through Project Development Facility.

d. To arrange training and visits, etc. of agripreneurs in setting up identified agribusiness projects.

Eligible Persons

- Assistance under the sub-scheme will be available to Individuals; Farmers; Producer.
- Groups; Partnership/Proprietary Firms; Self Help Groups; Companies; Agripreneurs; units in agri-export zones, and Agriculture graduates Individually or in groups for setting up agribusiness projects.
- For professional management and accountability, the groups have to preferably form into companies or producer companies under the relevant Act.

Eligible Financing Institutions

All Financial Institutions notified by the Reserve Bank of India where the ownership of the Central/State Government is more than 50% such as nationalized banks, SBI & its Subsidiaries, IDBI, SIDBI, NABARD, NCDC, NEDFi, RRBs & State Financial Corporations.

Choudhary Charan Singh National Institute of Agriculture Marketing (NIAM)

a. To undertake and promote the study of applied and operational research in problem areas of agricultural marketing and to act as a national level nodal point for coordination of various research studies and dissemination of technologies relevant to agricultural marketing in the country.

b. To impart training to various levels of personnel of organizations involved in agricultural marketing activities

c. To offer consultancy services to State and Central Departments, public-sector undertakings, co-operatives, etc. in the formulation of projects and prepare Master Plans for States, Export Institutions, Traders and Farmers.

Salient Features

a) Capacity Building/Training

a) Target Oriented Course Module

b) Demand driven Training

c) Developing linkages with the Stakeholders

b) Research Activities

NIAM will undertake research studies with focus on:

a) Detailed study/documentation of each innovation and intervention in agrimarketing sector for the purpose of policy formulation and replication.

b) Research for development of supply chain, Integrated value Chains for leading fruits and vegetables in collaboration with some national/ international university.

c) Education: The education mandate of NIAM aims at developing human resources by providing long-term structured courses in agricultural marketing through diploma courses.

d) Policy Advocacy: NIAM is pro-active in advising policy interventions and send a monthly advisory on various topics to DAC.

1. e-NAM

The electronic trading platform for National agriculture market was launched on April 16, 2016 in 21 Mandis across 8 States pilot trading of 24 commodities namely Apples, Potato Onion, Green Peas, Mahua Flower, Arhar whole (Red Gram), Moong Whole (green gram), Masoor whole (lentil), Urad whole (black gram), Wheat, Maize, Chana whole, Bajra, Barley, Jowar, Paddy, Castor Seed, Mustard Seed, Soya bean, Ground nut, Cotton, Cumin, Red Chillies and Turmeric.

Total of 585 mandis have been covered; 85% of these have assaying labs. The target for the number of mandis to be covered under eNAM in the first two years in 585 mandis across 16 States and two Union Territories was achieved in March 2018.

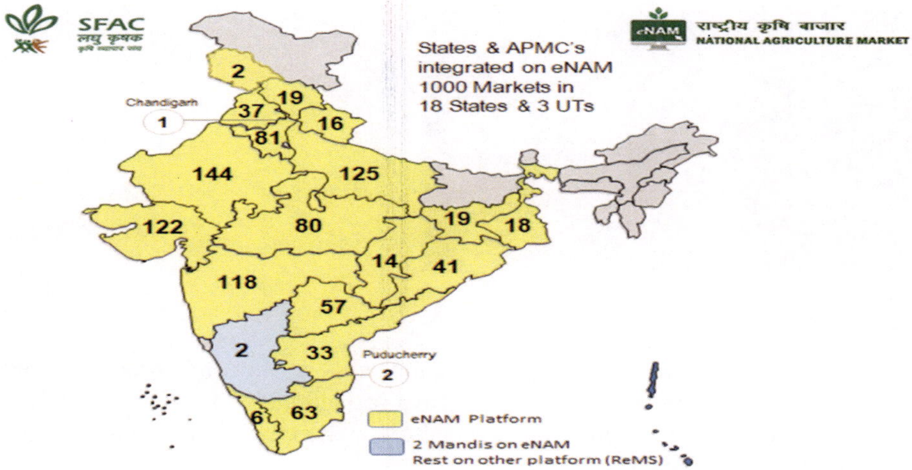

Source: https://www.enam.gov.in/

Success Stories

Case Study 1

- M/s. Nandagokula Industries was established in the year 2012 in Kundapura Taluka in Udupi district of Karnataka.
- The installed capacity of the unit was 1.4 MT/day and was operating at 78 per cent of its capacity.
- The operations in the unit were taken up both manually and automatically through machines.

- The processing unit received an amount of Rs.7.84 lakh under the VCA scheme 2014.
- The assistance acted as a supplementary source of fund for establishment of the unit. The VCA scheme has helped in establishment of the unit and been able to scale an appreciable turnover of Rs.1524.83 lakh in 2016-17.
- The unit is running in profit with gross profit of Rs.196.83 lakh in 2016-17.
- The unit has helped around 12 farmers (covering a cashew catchment area of approximately 60 acres) by procuring their cashew nuts.
- The unit employed as many as 80 labour directly round year in various activities of shelling, cutting, sorting, grading and packing.

Case Study 2

- MuskuVidyasagar, Resident of velkatoor village, Balkonda Mandal, Aged 45 years old. He owns Seven (7) acres of land and is having 20 years of experience in agriculture and agriculture marketing. He cultivates Paddy, Maize, and Soybean crops.
- Sri.Vidyasagar has been a seller in the market yard Nizamabad for last 15 years and all through, he was made to perform sale activity through Commission Agents.
- He has been a witness to open manual auction, Electronic, eNAM and e NAM (Direct Purchase) modes of selling in the market yard.
- Further, as he narrated that "he never had access to prices of commodities of other markets" and he used to stay back in the market yard for two to three days till completion weighment and wait 20 to 30 days for receiving his payment as there was total lack of transparency in price realization and deductions made by commission agents.

- Vidyasagar sold 26.16 quintals of soya white in eNAM (Direct Purchase centre) saved an amount of Rs.1427.00 through additional commission and Rs.1501.00 by way of reduced hamali charges. On a quantity of 26.16 quintals, he received a benefit of an amount of Rs.2929.00 in a single transaction of produce worth Rs 70,000.00(approx).
- Thus, by selling away his 270 quintals of Soya,Vidyasagar has become richer by more than Rs 30,000.00 off his agri produce in 2016, which otherwise would have gone to fill coffers of Commission agents. Further, he received the payment for the produce within 24 hours of selling through online transaction.

Case Study 3

- Shivam Cotton Industries, a cotton ginning unit is established at Dist. Jodhpur in Rajasthan which is known for its cotton belt.
- The project technology is latest & eco- friendly. The project has been set up with total project cost of Rs. 348.00 Lacs and credit facilities have been sanctioned by Corporation Bank, Chopasani Road Branch, Jodhpur.
- The firm has also availed assistance of Rs. 29.43 Lacs under Venture Capital Assistance Scheme of SFAC (Interest free Loan).
- The unit is working at the capacity of 60 MT/ per day during season and has achieved the turnover of Rs. 3572.57 Lakh in the year 2016-17 as against Rs. 1472.09 Lakh.

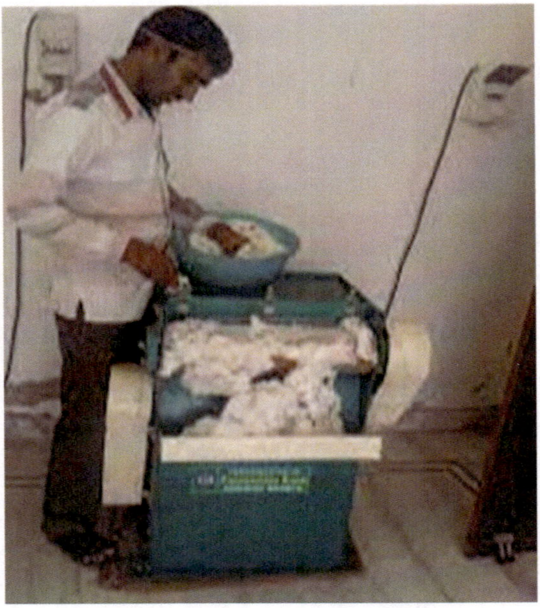

Suggestions

1. There should be a central agency for implementation, monitoring, and evaluations of ISAM.
2. Setting up Agricultural Mandis in every district to promote direct marketing of farm produce.
3. Involvement of private players in data collection and maintenance of database.
4. PPP options should be explored to bring expertise and value addition in marketing research and information.
5. NIAM should work in close coordination with SAUs and agriculture department for proper training of farmers, functionaries, entrepreneurs and other stake holders.
6. Evaluation of the approved project should be done in scientific manner and shortcomings (if any) should be properly addressed.

Conclusion

The GOI has been working continuously for the betterment pf farmers. Better price realization of farm produce will ensure the upliftment of farmers. The agricultural marketing plays a pivotal role in increasing and stabilizing farm income. With aim to improve the existing marketing infrastructure the Govt of India launched ISAM in 2013 by merging all the centrally sponsored schemes in agricultural marketing. This integration proved to be a major policy milestone achieved i n recent years in this field. But the critical question about sustainability of the scheme will depend on many factors like its implementation, monitoring, evaluation, budgetary allocations. Modernization of infrastructure, updating of policy framework regarding certification, maintaining export standards of produce are some of the key points which need to be addressed for the scheme to sustain in coming future.

References

http://sfacindia.com/

https://www.enam.gov.in/

https://www.nic.in/

Jabeen, A. and Jabeen, S. 2021. Agricultural Marketing in India – An Overview. International Journal of Agriculture and Rural Economic Research (ARER), 9(11): 6-10.

Kumar, M. 2018. Problems of Agriculture Marketing in India. International Journal of Research and Analytical Reviews, 5(4): 296-299.

Venkatesh, P., Singh, D. R., Sangeetha, V., Balasubramanian M, and Jha, G. K. 2021. Thechanging structure of agricultural marketing in India: a state-level analysis of e-NAM. Agricultural Economics Research Review, 34: 97-109.

www.agricoop.nic.in

16

Zoonoses Disability Adjusted Life Year (zDALY): Indicators for Measurement of Impact of Losses to Zoonotic Diseases

Arvind Kumar, Harsh Kumar Sharma, Rashmi Kailu Atul Anand, Zuhaib Fayaz Bhat, Sunil Kumar and Anil Bhat

Sher-e-Kashmir University of Agricultural Sciences and Technology, Jammu Jammu and Kashmir, India

Abstract

Zoonoses are those diseases that are transmitted from animal to human being and vive-versa. It occur at the interface of human and animal disease and partly because environment being common and their impact and management fall across two sectors they are also neglected. The Global Burden of Disease captures the impact of zoonoses on human health in terms of disability-adjusted life years (DALYs). Based on this, we estimate that in low income countries, zoonoses and diseases which recently emerged from animals make up more than twenty five percent of the DALYs lost to infectious disease and ten percent of the total DALYs lost. In contrast, in high income countries, zoonoses and diseases recently which emerged from animals represent less than one percent of DALYs lost to infectious disease and only fractionate percentage of the total disease burden. The costs of zoonoses and emerging disease to human, animal and ecosystem, and the health in terms of cost of treatment, cost of prevention, health burden and intangible and opportunity costs. The eco-health concepts of trans-disciplinarity, participation and equity can help in assessing the importance of zoonoses in developing countries and illustrate these with an example of assessing milk-borne disease.The burden of human diseases in populations, or for an individual, is frequently estimated in terms of one of a number of Health Adjusted Life Years (HALYs). The Disability Adjusted Life Year (DALY) is a widely accepted HALY metric and is used by the World Health Organization and the Global Burden of Disease studies. Many human diseases are of animal origin and often cause ill health and production losses in domestic animals. The economic losses due to disease in animals are usually estimated in monetary terms, being commercial in nature. The monetary impact on animal health is not compatible with HALY approaches used to measure the impact on human health. The societal burden of zoonotic diseases that have substantial human and animal disease burden we propose methodology which can be accommodated within the DALY framework. Monetary losses due to the animal disease component of a zoonotic disease can be converted to an equivalent metric using a local gross national income per capita deflator. This essentially gives animal production losses a time trade-off for human life years. This is the time required to earn the income needed to replace that financial loss. This can then be assigned a DALY equivalent, termed animal loss equivalents (ALE), and added to the DALY associated with human ill health to give a modified DALY. This is referred to as the "zDALY". ALEs could also be estimated using willingness-topay for animal health or survey tools to estimate the replacement time value for animals with high societal or emotional value (for example pets) that cannot be calculated directly using monetary worth. Thus the zDALY estimates the impact of a zoonotic disease to animal and human health. The losses due to the animal disease component of the modified DALY are straightforward to calculate. A number of worked examples such as echinococcosis, brucellosis, Q fever and cysticercosis from a diverse spectrum of countries with different levels of economic development illustrate the use of the zDALY indicator.

Introduction

The quest for data and analysis of the economics of zoonoses control is now in its fifth decade: The World Health Organization (WHO) Executive Board having passed a resolution, recommending that a group of experts meet and discuss the subject and commission some pilot studies. (a) The health economics recognised as a sub-discipline of economics (b), the cost–benefit analysis established a notion for long enough to merit the writing of a survey article (c) and the difficulties of putting a monetary value on human life (d) the alternative non-monetary approaches being developed. The development of veterinary or animal health economics, partly spurred on by the gathering of experts initiated by that same WHO resolution. The International Society for Veterinary Epidemiology and Economics held its first meeting marking the formal launching of the fraternity.

An understanding of the economic and social impacts of diseases is core to the decision making process for disease control. The animal disease impact is widely studied and experimented using economic models on the basis of monetary costs. Thus, the animal losses can be estimated for the values of both livestock and livestock products lost through diseases. In health economics, the cost of human disease can be quantified in terms of cost of treatment and loss of income during convalescence are also included in cost of illness studies. The calculation of a monetary burden of human ill health, to incorporate death and disability as well as the cost of illness, has been criticised for a innumerable reasons. There is no standard methodology to calculate the monetary value of life or loss Indeed that has occurred, this has been modelled to estimate the statistical value of life with gross domestic product per capita as a covariate [8]. Likewise, the costs of medical treatment as well as loss of wages are usually higher in upper than in lower income countries. This could perversely evident that the greatest economic impact of disease is in wealthy countries, despite having higher life expectancies and lower disease incidence than low income countries. Thus estimations of the monetary cost of disease effectively assign higher values to equivalent health outcomes in developed countries than to the same outcomes in the developing countries. Because of these and the wider issues around valuing human life, non-monetary population health metrics such as disability adjusted life years (DALYs) have been developed to quantify the burden of premature death and disease and injury. DALYs are calculated by adding the number of years lived with disability adjusted for the severity of the disease (YLDs) and the number of years of life lost due to premature mortality (YLLs) [11]. DALYs have been adopted by the WHO to quantify the global burden of disease [12] and the global burden of foodborne diseases. They can also be used in cost-effectiveness studies such as cost per DALY averted. In the animal health field, there have been attempts to mirror these burden estimates for animal diseases. There are important conceptual difficulties: how to compare within and across animal species rearing under different climatic zones, the fact that for livestock producers maximising their animals' life expectancy is not necessarily a target and that the same physical disabilities have completely different results in different livestock productions systems. The work sponsored by the World Bank assigned livestock unit values to various animal species, similar to the proven and verified approach based on animal profiles and mapped the total units to be lost through death, destruction or slaughter. Despite emphasizing the unevenness, bizarre differencesand major gaps in the data available on livestock disease impact, its illustrators to calculate the impact the relative geographical burden of fatal animal diseases. By contrast, the well-established and proven methods of animal health economics, allow for the impacts of burden of animal disease, including commercial trade effects and the cost of mitigating measures to reduce various kinds of losses, to be estimated in livestock across production systems and regions, and aggregated into a single currency for universal unanimity. The willingness-to-pay and other household survey methods can be well used, the monetary value of companion animals and wildlife can be well-estimated in a range of contexts. As well as owners' monetary valuations, financial awards in legal cases are often linked to the lifetime cost of keeping companion animals and for wildlife, other monetary considerations include their commercial contribution. As a non-monetary value, DALYs do not capture other societal impact of burdens that may not be direct involvement of human and health-relatedsubjective. With the zoonotic diseases there is clearly potential for a human disease and an animal disease burden, as by definition zoonoses infect both animals and humans thereby affecting both species equally. There has been no satisfactory metric developed that can incorporate both these burdens and estimate their relative share precisely and accurately in the societal cost of disease. For zoonoses control projects, cost sharing in proportion to expected disease control benefits has been projected and proposed, but this is related only to the estimated monetary benefits in terms of livestock disease prevention and monetary savings from reduced cost of human as well as livestock illness, which are assessed separately from the DALY component. A range of health-adjusted life year (HALY) measures have been developed, of which the two currently most commonly used are the quality-adjusted life year (QALY) and the disability-adjusted life year (DALY). The differences between the DALY and the QALY are explained in the context of a zoonotic disease. The DALY measures burden of disease, and disease control 'averts' DALYs. A DALY value of 0 is equivalent to a perfectly healthy life year and 1 to death, with values in between representing different levels of disability. The DALY has two components: years of life lost (YLL: measuring the impact of death) and years of life lived with disability (YLD: measuring time spent living with health problems). The DALY has been widely adopted, notably by WHO, as an international measure of the impact of disease and as a basis for assessing the global burden of disease impacting both livestock as well as humans. As far as the concept of One Health, the DALY has become the preferred measure for expressing the impact of zoonoses on human health. The DALY, in its global application also packs a moral punch as it implies that across the world a life is a life, no matter whose it is nor where that person lives or what they earn. Expenses on treatment and prevention in the human sector in human health economics, costs are divided into direct medical and non-medical costs as the latter including healthsector overheads and costs that patients necessarily incur, from all aspects such as transport and hired care-giving) and indirect costs viz., the patient costs while ill: the time off from work, the seeking treatment, and the hired household help. The various direct and indirect costs may or may not be taken into account whether the discussion is from the viewpoint of an individual stakeholder or a small ethnic group of society as a whole.

Animal health losses are the losses due to illness in livestock are valued in terms of their monetary impact on livestock output. This is defined as the total goods produced whether for home consumption or sale of raw produces viz., meat, milk, skins, wool, animal traction, manure etc. and other by-products, minus animals purchased or brought in, plus any change in herd value occurred in the meantime. A wide range of time periods, methods and scales are used say for example one-year farm budgets, herd models using production parameters such as death and fertility rates to calculate output, and analyses of the impact of largescale changes in the health status of livestock populations on supply, demand and prices (producer and consumer surplus estimates). As time progresses, due to increasing web of complexities, the wider impacts may also need to be quantified. The possible impacts on commercials include a fall in prices and a deduction in the number of marketing avenues and outlets as a result of movement controls or import bans or the COVID pandemics. The endemic diseases such as region wise can restrict the various farm breeds farmed and the type of livestock production systems used. Within the livestock sector, the animals have clear commercial values; however, when considering companion animals and wildlife, estimation is more complex – involving techniques that are also familiar in the human health economics sector, such as willingness to pay and so on so forth. The expenses on treatment and prevention in the livestock sector as well as suffering losses as a result of their animals' illness, livestock keepers and the local/national veterinary and municipal services incur expenses as they attempt to mitigate or prevent these losses. The commercial trade-off between accepting losses in livestock and spending money privately or publicly on improving their health standards has been termed the 'loss–expenditure' frontier and analysed using the tools of economic theory.

In order to be incorporated into the DALY metric, animal and human deaths and their production losses, need to be quantified into an animal and human disease burden metric system which can stand alongside the estimation of DALY. Such a metric should indicative of the impact of the human and animal disease in terms of the time that might be required to recoup the losses caused by its illness. These impacts can thus be viewed and interpreted as time that the livestock owner has been forced to sacrifice to replace the loss of his animal or to make up for the reduced and low output that a non-fatal animal disease causes. A modified DALY for zoonotic disease, termed zDALY, that has an additional component, termed animal loss equivalents (ALE). The ALE can be estimated from the monetary value of livestock losses and local per capita income by using a time commercial trade-off approach to estimate an equivalent impact burden measure to the human population. The defined rationale is therefore to include animal health losses into the DALY framework to assess the burden of diseases that transmit between animals and humans. In the DALY framework, non-fatal human illnesses and injuries are given a disability weight (DW) and the duration of this DW gives an equivalent number of lives lost and is essentially a commercial time trade-off by assigning the time lived with the disability to a shorter time of healthy years or lived without disability). The ALE component in the zDALY is also given a time equivalent of healthy years.

One of these priorities was the development of a comprehensive methodology for calculating the societal burden and impact of disease attributable to zoonoses. The zDALY metric provides a straight forward, practical and rigorous and valid answer to this need and the question arisen to assess and measure. The decision-policy makers increasingly require comprehensive economic metrics traits summarising and comparing the benefits and costs of controlling zoonotic diseases. The impact of burden of disease in people is conventionally quantified in non-monetary terms, usually a disability-adjusted life year (DALY), whereas the losses due to disease in livestock, are quantified in monetary terms. The potential for the development of a non-monetary metric for ill health in animals, based on life years lost and disability, is discussed and rejected. Within and across animal species and livestock production systems, maximising life spans is not a consistent target and morbidity/disabilities have very different weights and often lead to culling and slaughtering. By relating livestock losses to a measure of national income forgone, the recently developed alternative of converting monetary losses due to livestock illness into an animal loss equivalent (ALE) provides a viable and valid solution. Based on this, the economics of controlling zoonoses is reanalysed and four options for quantifying and comparing benefits and costs are examined and illustrated using numerical examples. These are i) the simplistic grouping of all monetary elements and their comparison to DALYs averted (described as the aggregate net cost method), ii) the separable costs method, iii) the use of ALEs to convert all benefits to a non-monetary equivalent, termed the zoonotic DALY (zDALY), or iv) the use of a full monetary cost-benefit analysis, based on converting DALYs to a monetary equivalent. The strengths and weaknesses, opportunity and threats of each are discussed. For effective prioritisation and decision-making, it is vital that an analytical approach is widely adopted which yields consistent results and supports the control of menace of zoonoses.

Analysing the economics of interventions to control zoonoses following are the hows the costs of disease provide the basis for estimating the impact of a particular disease control intervention in terms of reductions in some or all of the four categories. Adding up all of these benefits, from both the human and animal health sectors, yields the total societal benefit from controlling the disease using that intervention. These advantages are then compared to the expenses, which for a zoonotic disease being implemented simultaneously by the human faculty and the livestock faculty; for example, in the case of rabies, the former provides post-exposure prophylaxis for humans and the latter deals with vaccination of dogs. Basic conventions in human and animal health economics In human health economics the three main approaches used for assessing impacts are cost–benefit analysis (CBA), cost-effectiveness analysis (CEA) and cost–utility analysis (CUA). CBA compares expenses to monetary benefits and CEA compares expenses to a measure of effectiveness – which can range from number of illness successfully treated and to lives saved. Increasingly, CEA refers to the expenses of measures taken per DALY averted, whereas CUA refers to the expenses per unit of utility, a measure of welfare as defined in economy theory, and thus usually per QALY gained. In the international scenario and in zoonoses control, the most frequently cited measure is a cost-effectiveness ratio (CER) based on expenses per DALY averted. On all occasions, costs are the numerator, so that strategies yielding smaller results like lower cost–benefit, cost-effectiveness or cost–utility ratios can be preferred. When comparing advantages and expenses, in human health the customary is to evaluate the expenses and impacts of a range ofmeasures to tackle a particular health problem. In every case, the extra advantages from an additional measures taken are compared to the extra costs, thus resulting an incremental cost-effectiveness ratio (ICER). At an international level, until recently, measures of expenses less than twenty five united state dollars per DALY were regarded as highly attractive and more than seventy five dollars as attractive. Current WHO cost-effectiveness thresholds define measures costing less than one year's gross domestic product per person as being very cost-effective, those costing between one and three years ofgross domestic product per person as simple cost-effective, and those costing more than three years ofgross domestic product per person as not cost-effective. The animal health economics also uses CBA. The relating costs to measures of effectiveness, such as a lowering in disease occurance or the number of livestock get prophylaxis, though it has been recommended that animal sector economics utilizes more of CEA. Both at the on-field farm level and at the livestock service level, break-even evaluationprove useful role when results are uncertain, as it implies to what minimum level of benefit or return is required to cover loss expenses. In livestock health economics, the basic parameters for identifying costs and benefits is the 'partial evaluation' framework. Costs are divided into the 'extra costs' of the intervention and any extra inputs required for a healthier or more numerous livestock population plus any 'revenue forgone' like negative side-effects of the any implications. Benefits consist of 'extra revenue' like the reflection a lowering in disease losses and thereby 'costs saved'. This outline is generally applied at a farm level, but implies a basis for recognizingcost advantages of any interventions made on any level or scale. The emphasis on extra revenue and forgone revenue saved recognised as an incremental evaluation analogous to the ICER, in which the profit unit is monetary rather than a measure of effectiveness. These differences in analytical focus and criteria are a clear reflection of the ways in which the human and livestock health faculties operate. Human health and the control of disease in human is considered a public good overwhelmingly, both for communicable and non-communicable diseases, elective interventions such as cosmetic treatments or prolonged fertility treatments. In livestock health, the prevention and control of highly contagious, transboundary diseases and zoonoses is taken a public good, whereas endemic diseases which affect livestock productivity and farmers incomes like mastitis or non-zoonotic worm infestations implies to livestock owners. Hence, much of livestock health economics targets on farm-level decisionmaking and on monetary benefits. In human beings health, the target is on spending public-sector money in such a manner as to reach to maximum health impact, so a complete range of options areexpensed out and their after effects are compared. Four options for zoonosesin order to assess the economics of interventions for prevention andzoonoses are, the impacts on the human and livestock health faculties need to be conglomerated so that decision-makers in both sectors can evaluate, analyse and interpret results in a way which is more meaningful and applicable both to their sector and to society. Four options can be considered for implications are, taking three neglected zoonoses as: at national level, one at county level and third at city level. The intervention consisted of enhanced investments in all works which were already in progress, so the additional expenses are considered as incremental costs, but the priorexpenses are not considered as costs saved or reduced expenses. Hence, the only advantage category represented is DALYs averted. Throughout, costs are, applicable at the time the study was undertaken, in order to retain the originality between the livestock losses, disease control expenses and GDP. The aggregate net cost per DALY averted,the simplest and most evident way to compare all the benefits and expenses would be to combine all the monetary components viz., intervention costs, reductions in health cost and livestock losses and compare them to the single nonfinancial unit, the lowered impact of burden of human disease in DALYs as expressed averted. In order to obtain a net cost per DALY averted, the advantages are subtracted from the expenses, and divided by the DALYs. Using the customary of human health economics, this method does gives us a series of costs per DALY averted, which can be ranked, with rabies prevention and control being the most expensive, and brucellosis prevention and control being the least. But, for all the examples except rabies, the cost per DALY averted is negative. This is because the substantial monetary benefits from costs saved and from the animal sector outweigh the costs before even considering the reduction in the human health burden of the said disease. The human health sector interventions, the implicit approach is of net cost, since the overall enhancement in cost is compared to the additional DALYs averted. The separable costs method aims to apportion expenses for controlling the disease in proportion to the monetary benefits obtained from its prevention and control. This type of expense allocation is the classic approach used for interventions involving several faculties. Although it is usually applied to fully monetised calculations of advantages and expenses, it has been adapted for zoonosesprevention and control. Having quantified all the critical control elements, the first step is to work out the share of monetary advantages accruing to the human health faculty and the share accruing to the livestock health faculty. Then, these percentages are applied to the costs, to derive the share of costs which should be attributed to each faculties, where CS stands for 'cost share'. The cost-effectiveness ratio (CER) for human health is then counted as its share of costs divided by the DALYs averted. For the animal sector the monetary advantages are divided by the faculty's share of costs. This yields a benefit–cost ratio (BCR) which is identical to that which would be obtained for the human sector if the expenditure saved were divided by its expense share, so without considering the DALYs. These are credible avenues which make good sense to health service providers. This method has been widely adopted for zoonosesprevention and control interventions program [19, 20]. By emphasising that a reasonable share of monetary advantages accrues to the human health sector in terms of savings in treating people, it seeks to promote greater human health sector funding for disease prevention and control, even when the control program like vaccinating reservoir animals as for rabies or brucellosis, is mainly undertaken by the livestock health sector. As the example of rabiesexplains, if there are no monetary advantages to either sector from the application and intervention, but DALYs are averted, all costs are attributed to the human health faculty, so that result will be the same as for the overall net cost method. As cost–benefit analysis, within human health economics, a monetary CBA is sometimes undertaken, using various methods to value human life and suffering. Thus, RBH is calculated and evaluated in monetary terms as GDP × DALYs and a BCR deduced. A net benefit (sum of benefits less sum of costs) can also be calculated similarly. For a multi-year project, where the values are discounted to their present value, this would be in the form of a net present value (NPV), and an internal rate of return (IRR) could usually also be evaluated. Two things are apprehended from this fully monetised evaluation are, first, including the value of the human health advantages in the BCRs makes these much larger than they were before. Second, their inclusion increases the share of benefits attributed to human health, for all zoonotic diseases. For the rabies example, which consists of increasing investment in both sectors, the only measured benefit is in DALYs, so only a fully monetised CBA allows for a BCR to be evaluated, yielding a value of two. zDALY analysis: incorporating the ALE lastly, although, the ALE is a non-monetary metric representing the cost of livestock disease using a measure of national income, allows for the CBA process above to be undertaken in reverse, by converting monetary units into non-monetary units. Thus, all monetary advantages are converted to income equivalents by dividing their fiscal amount by the monetary value of one year's per capita of GDP. For livestock losses, this results an ALE. For human health monetary costs saved, a similar process results a human monetary loss equivalent or HLE. The difference in enhancement as the share of human benefits increases, a function of inclusion, as the DALYs averted in the equation. As the same value of one year's GDP used in this evaluation as for the BCR calculation, the shares of each faculty in total benefits are identical for these two approaches, but as noted above, higher for human health than when using the separable costs method. The CERs produced can be ranked and, expressed as a proportion of one year's per capita of GDP, given, thus ranging well within WHO's threshold of very cost-effective implications and interventions to calculate the impact measurements.

Conclusions

A standardised and proven approach to measure results of zoonoses prevention and control in an integrated way, the economic efficiency of such prevention and control measures could be compared internationally and the best approach selected to measure the impact of zoonotic diseases. The ALE, as a way of converting monetary livestock benefits into a non-monetary metric, enables livestock health benefits to be integrated into human health economic analyses. This is a realistic,reliable and viable option to calculate the impact of zoonotic disease to gross domestic products. The greatest strength of the zDALY has been in analysing and calculating global health burden of zoonotic diseases, because mortality from all causes need to add up to total deaths, although there are still unresolved issues of doublecounting avoidancefor co-morbidities. But, in livestock health, for many diseases and production systems there is no consensus on what overall losses are from specified zoonoses. The zoonotic disease losses are assessed one disease at a time, so that the sum of deaths ascribed to various individual causes often greatly exceeds the total number of deaths observed. Hence, a huge gap sustains to fill in terms of providing more coherent data on the impacts of livestock diseases. Conversely, converting the zDALY into monetary terms based on GDP does detract from its appeal as a universal metric. The WHO's value-for-money criteria implicitly already accept this and comparisons are made not an individual, but on a country, income basis. The issues arise when implicating regional or global life expectancies for zDALY calculations, similar implications are taken into consideration. The fiscal conversions do not diminish the value of being able to calculate and compare the burden of zoonotic diseases across countries in terms of zDALYs.

References

A.A. Sultanov, S.K. Abdrakhmanov, A.M. Abdybekova, B.S. Karatayev, and P.R. Torgerson (2016). Rabies in Kazakhstan, PLoSNegl. Trop. Dis. 10: e0004889.

A.H. Havelaar, M.D. Kirk, P.R. Torgerson, H.J. Gibb, T. Hald, and R.J. Lake. (2015). World health organization global estimates and regional comparisons of the burden of foodborne disease in 2010, PLoS Med., 12: e1001923.

A.P.M. Shaw, G. Cecchi, G.R.W. Wint, R.C. Mattioli, and T.P. Robinson.(2014).Mapping the economic benefits to livestock keepers from intervening against bovine trypanosomosis in Eastern Africa, Prev. Vet. Med., 113: 197-210,

B. Devleesschauwer, A.H. Havelaar, C. Maertens de Noordhout, J.A. Haagsma, N. Praet, P. Dorny. (2014). Daly calculation in practice: a stepwise approach, Int. J. Public Health., 59: 571-574.

C. Benner, H. Carabin, L.P. Sanchez-Serrano, C.M. Budke, and D. Carmena.(2010). Analysis of the economic impact of cystic echinococcosis in Spain, Bull. WHO, 88: 49-57.

C. Milligan, A. Kopp, S. Dahdah, and J. Montufar. (2014). Value of a statistical life in road safety: a benefit-transfer function with risk-analysis guidance based on developing country data, Accid. Anal. Prev., 71: 236–247.

C. Stein, T. Kuchenmüller, S. Hendrickx, A. Prüss-Ustün, L. Wolfson, and D. Engels. (2007). The global burden of disease assessments—WHO is responsible? PLoSNegl. Trop. Dis., 1: e161.

C.J. Murray. (1994).Quantifying the burden of disease: the technical basis for disabilityadjusted life years, Bull. World Health Organ., 72: 429-445.

C.J.L. Murray, J.A. Salomon, and C. Mathers. (2000).A critical examination of summary measures of population health, Bull. WHO, 78: 981-994.

D. Rice. (1994). Cost-of-illness studies: fact or fiction? Lancet, 334:1519–1520.

E.J. Neumann, J.B. Kliebenstein, C.D. Johnson, J.W. Mabry, and E.J. Bush. (2005). Assessment of the economic impact of porcine reproductive and respiratory syndrome on swine production in the United States, J. Am. Vet. Med. Assoc., 227: 385–392.

F. Roth, J. Zinsstag, D. Orkhon, G. Chimed-Ochir, G. Hutton, and O. Cosivi, (2003). Human health benefits from livestock vaccination for brucellosis: case study, Bull. WHO, 81: 867–876.

J.A. Haagsma, A.H. Havelaar, B.M. Janssen, and G.J. Bonsel (2008). Disability adjusted life years and minimal disease: application of a preference-based relevance criterion to rank enteric pathogens, Popul. Health. Metr., 6: 7-14.

J.A. Haagsma, S. Polinder, A. Cassini, E. Colzani, A.H., andHavelaar. (2014). Review of disability weight studies: comparison of methodological choices and values, Popul. Health. Metr., 12:20-28.

J.A. Rushton. (2009). Review of the application of economics to animal diseases and health Problems, in: J. Rushton (Ed.), The Economics of Animal Health and Production, CABI, Wallingford, UK, pp. 194-197.

J.A. Salomon, T. Vos, D.R. Hogan, M. Gagnon, M. Naghavi, and A. Mokdad.(2012). Common values in assessing health outcomes from disease and injury: disability weights measurement study for the Global Burden of Disease Study, Lancet, 380:2129–2143.

M. FasihiHarandi, C.M. Budke, and S. Rostami. (2012). The monetary burden of cystic echinococcosis in Iran, PLoSNegl. Trop. Dis. 6: e1915.

M.J. Counotte, G. Minbaeva, J. Usubalieva, K. Abdykerimov, and P.R. Torgerson.(2016). The burden of Zoonoses in Kyrgyzstan: a systematic review, PLoSNegl. Trop. Dis.10: e0004831.

P.L. Moro, C.M. Budke, P.M. Schantz, J. Vasquez, S.J. Santivanez, and J. Villavicencio. (2011). Economic impact of cystic echinococcosis in Peru, PLoSNegl. Trop. Dis. 5: e1179.

P.R. Torgerson, and C.M. Budke. (2014). Echinococcosis – an international public health challenge, Res. Vet. Sci., 74: 191-202.

P.R. Torgerson, B. Devleesschauwer, N. Praet, N. Speybroeck, A.L. Willingham, and F. Kasuga. (2015). World Health Organization estimates of the global and regional disease burden of 11 foodborne parasitic diseases, 2010: a data synthesis, PLoS Med., 12: e1001920.

P.R. Torgerson, P.M. Dowling, and M.N. Abo-Shehada. (2001). Estimating the economic effects of cystic echinococcosis. Part 3: Jordan, a developing country with lower-middle income, Ann. Trop. Med. Parasitol. 95:595–603.

R. Bennett. (2003). The "direct costs" of livestock disease: the development of a system of models for the analysis of 30 endemic livestock diseases in Great Britain, J. Agric. Econ., 54: 55–71.

R.J. Brooke, A. Van Lier, G.A. Donker, W. Van der Hoek, and M.E.E. Kretschmar. (2014). Comparing the impact of two concurrent infectious disease outbreaks on the Netherlands populations, 2009, using disability-adjusted life years, Epidemiol. Infect., 142 :2412-2421.

S. Henrard, B. Devleesschauwer, P. Beutels, M. Callens, F.D. Smet, and C. Hermans.(2014). The health and economic burden of haemophilia in Belgium: a rare, expensive and challenging disease, Orphanet. J. Rare. Dis., 9: 39-45.

S.R. Palmer, L. Soulsby, P.R. Torgerson, and D.W.G. Brown. (2011). Oxford Textbook of Zoonoses Biology, Clinical Practice, and Public Health Control, Oxford University Press, Oxford, England.

T.J.D. Knight-Jones, and J. Rushton (2013). The economic impacts of foot and mouth disease - what are they, how big are they and where do they occur? Prev. Vet. Med., 112: 161-173..

17

Impact of Climate Change on Agriculture in the North Himalayan Region

[1]Ankita Kumari and [2]Pavnesh Kumar

[1]Department of Management Sciences
[1,2]Mahatma Gandhi Central University Motihari, Bihar
[2]Professor, Department of Commerce and Management Sciences

Abstract

Uttarakhand is bestowed with several rivers, tourists' destinations, pilgrimage sites, and suitable climatic conditions for growing fruits, vegetables, food grains, rearing livestock, tea garden practices, etc. On the other hand, the lack of modern technology, innovation in the agricultural system and people's reluctance to adopt it could not take place in economic development properly. Furthermore, growth and development could not take place as a part of the Uttar Pradesh state due to its harsh climatic conditions, rigorous terrain, and distinct identity.The state is believed to be one of the poorer states today. This area lacked adequate infrastructure due to its inaccessibility. The idealagrarian and geographic circumstances might be equally applied to the developmental processes. The area as a whole is vulnerable ecologically. The state's overall scenario is further changed by the diverse socio-economic activities, strict traditional beliefs, and hardworking potential. Only the need for the hours is to formulate and put into action rational policies and prepare for the long-term sustainability of the state. In order to achieve economic growth, it is imperative that policies, plan, and beliefs be drastically changed. This paper aims to review the impact of climate change on agriculture in the North Himalayan Region, Uttarakhand.

Keywords: Agricultural system, Climatic condition, Economic development, Infrastructure, Natural resource, Socio-economic condition.

1. Introduction

Uttarakhand is extremely vulnerableto climate-related changes due to its varied geography and rich glacier mass. The relationship between climate change and agriculture is not a new phenomenon. In the last 50 years, there has been a lot of academic focus on dealing with climate change issues in agriculture. Climate change is believed to affect agriculture by inducing changes in farmer's behaviour, quantity, quality, cost of production, changes in production, consumption, prices & trade patterns; changes in market responses at global & local levels and these changes not only depend on the domestic and international adaptive capacity their economic impacts also vary by region, by sector, and by stakeholders' groups. Climate change is a major global environmental problem and also an issue of great concern to all countries, irrespective of their size or level of development. It is established that adverse impacts are likely from increased frequency of extreme weather, floods & droughts, and submergence of coastal areas due to sea level rise and extreme climate variability. It is considered that mountain regions are vulnerable to climate change & it will have direct impacts on livelihoods as most of the economic & livelihood sectors are vulnerable to the effects of climate change.

As per report "The increasing intensity of rainfall events can resultin floods, causing crop losses andimpeding access to markets to sell agricultural goods." Agricultural producers are substantially impacted by changes in ozone, greenhouse gases and climate change because agriculture and fisheries depend on specific climate conditions.Temperature changes which cause habitat ranges and crop planting dates to move and droughts and floods due to climate change may hinder farming practices.

A new study on the state in the middle of the Himalayan range by the Germany-based Potsdam institute for climate impact research (PIK) and the energy and resources institute (TERI) in New Delhi has forecasted the worst impacts will be in higher elevations. Existing pressures on farming include a decrease in land held per person, lack of irrigation infrastructure, crop depredation by animals like wild boars or monkeys, and a waning interest in farming among youth. This may accelerate the trend of people migrating and leaving land follow.

Present impacts of Climate Change on Uttarakhand's Residents

- Changing temperatures
- Upward moving snowlines
- Receding glaciers
- Erratic rainfall
- Reduction of snow in the water

- Changed cropping seasons
- Shifting cultivation zones for certain crops
- Drying up of perennial streams.

After reviewing the literature, it was found that Uttarkashi, Chamoli, Rudraprayag, and Pithoragarh, districts at higher elevations, are warming fastest. Almora, champawat, Puri Garhwal, and Tehri Garhwal are the districts likely to experience increased water stress. That is why climate change in Uttarakhand will increasingly force people to abandon forming at high altitudes and move to the plains over the next 30 years.

Uttarakhand, which covers an area bigger than costa Rica maybe 1.6-1.9 degrees Celsius warmer by 2050. Most people in Uttarakhand live in rural areas; 71% depend on rainfed agriculture, practiced through terrace farming on hill slopes.

Uttarakhand was formed on 9th November 2000 as the 27th state of India, when it was carved out of northern Uttar Pradesh. It is located at the foothills of the Himalayan Mountain ranges. It is primarily a hilly state, having international boundaries with China (Tibet) in the north & Nepal in the east. On its northwest lies Himachal Pradesh, while on the south is Uttar Pradesh. It is rich in natural resources, especially water and forests with many glaciers, rivers, dense forests & snow- clad mountain peaks, char- dhams, the four most sacred & revered Hindu temples of Badrinath, Kedarnath, Gangotri, and Yamunotri are settled in the mighty mountains.

Three ways of climate change may impact agriculture: -

a) Increased water stress
b) Increased risk of floods
c) Change in crop yields

The study indicates that increased water stress could further reduce crop yields under RCP 4.5 & RCP 8.5 scenarios, adding to existing migration pressures. There will be more frequent droughts, fewer rainy days but heavier rainfall on those days,and less rain or snow in winter.

RCP stands for 'representative concentration pathway.' Scientists use RCPs to study how different concentrations of greenhouse gases in the atmosphere will change because of human activities. RCP 8.5 assumes a very high level of future concentrations due to unsuccessful efforts to curb emissions; RCP 4.5 is based on a median level, likely to result in a 2 C – 3 C global temperature rise.

According to the rural development and migration commission, Uttarakhand, the inability to diversity livelihoods in rural areas is the most significant factor pushing out-migration (50%), followed by lack of educational institutions (15%) & health care facilities (9%). Because of migration, an increasing number of villages in Uttarakhand are uninhabited. According to a 2018 survey by the commission, 734 villages in the state have become uninhabited since 2011.These villages are spread across all 13 districts & are often referred to as ghost villages.

Agriculture also affects climate: -

- Agriculture is also the reason for deforestation & a change in land use from natural ecosystems that take up & store carbon dioxide (CO2) from the atmosphere to farmland.
- Greenhouse gases (GHGs) have different sources within agriculture.
- The livestock sector alone is responsible for about 44% of human-induced methane (CH4) emissions, 53% of nitrous oxide (N2O) & 5% of carbon dioxide (CO2).

As per the prior study, poverty is a growing phenomenon in most areas of the Uttarakhand State, especially in the mainland (hills parts), which constitute about 85%of the total geographical location. Negative impacts on agricultural productivity and loss in the quality and quantity of forests directly affectthe well-being of the rural households who depend on these two primary resources for their sustenance. Throughout the state, serious environmental problems have already emerged.

These environmental problems are particularly noticeable in the mainland of Uttarakhand as a formof degradation and depletion of forest resources. Landslides, landslips, cloudbursts, and flash floodsare the major threats toland stability (Sati1993). Agriculture is the main occupation of thepeople, and about 75% population is engaged with agriculture and its allied practices (includinghorticulture, herb production, and rearing of animals). Present conditions of resources as a form of natural vegetation, crops, horticultural farming, herbs, tea garden practices, livestock rearing, hydropower projects, and economic development of Uttarakhand.

Methodology

We have used secondary data in the form of published government report related to agriculture, journal, articles, periodicals and books.

Discussion

Representative farming situations in Uttarakhand Himalaya

Like most of India, agriculture is one of the significant sectors of the economy of Uttarakhand state. About 86% of the state consists of hills, but most of the croplands are in the Terai and Bhabar region plains, where improved agriculture is practiced. On the other hand, the hilly areas of Uttarakhand state (parts of north-western Himalayas) traditionally grow diverse crops as polycultures. The crops include coarse cereals (minor millets-finger millet, barnyard millet, foxtail millet, little millet, prosomillet), rice, wheat, barley, pulses (horse gram, black gram, black-seeded local soybean, lentil, cowpea, rice bean), oilseeds (mustard, sesame), several underutilized crops (amaranths, buckwheat, chenopods, tubers, yams, cucurbits, little leafy vegetables, perilla, and others), numerous medicinal herbs, etc. Most of these native crops and their traditional landraces form the "functional foods" to hill farming communities with nutraceutical properties and high nutritional value. The subsistence harvests and uses of wild resources by native communities have been other essential features of traditional hill farming.

Fruits like apples, oranges, pears, peaches, plums, and citrus are also grown and important to the food processing industry in certain areas. Livestock (buffalo, cattle, goats, sheep, and poultry) are integral to hill farming and contributes substantially to the rural economy. In Uttarakhand hills, a self-contained food system exists specific to different farming situations and communities deeply rooted in local traditions and culture. However, the forces of globalization and the vulnerabilities arising from poverty, discrimination, and marginalization in some sections of farmer households in the hill communities are increasingly affecting the native food culture. Limited access to diverse food resources is affecting the nutrition and health of native communities, causing malnutrition among children and women. Traditional hill agriculture is highly knowledge-based, and farmer-led innovations are predominantly practiced. All farmer-led innovations are socio-economic and are not technical. The driving forces for the emergence of the farmer-conducted innovation paradigms are constraints to crop and livestock production.

Non-availability of quality seeds of native crop varieties, lack of human resources (farm laborers), poor market infrastructure for native crops, high land fragmentation, and the changing climate, significantly more significant frequency and severity of droughts during the past two decades or so, have been the considerable constraints to traditional rainfed agriculture of Uttarakhand hills. Further, poor management of common property resources (CPRs) leads to declining fodder resources for livestock in crop-livestock small-scale mixed farming and diminishing sources of wild harvested foods. The dynamic relationship among crops, livestock, and CPRs is increasing.

Conclusion

Theprimary profession ofHimalayan people is primitive agriculture. Their economy is hence dependent on agriculture and related activities. Due to the remoteness, bad transportation, rugged terrain, and the people's lack of motivation to take action, the motivation of governments and non-governmental organisations as well as the deployment of modern technologies in agriculture did not result in an appreciable mode of development. It is important to find a timely solution to the concerns of soil erosion and landslides. Natural disasters are common in this state and happen every year. As a result, soil infertility, fragility, and instability are frequent occurrences. Due to its undulating topography, harsh climate, weak economy, and lack of development, this state is ecologically fragile. Natural resources are abundant, including forests, animals, water, herbs, and suitable climatic conditions, as was already indicated. However, up until now, they have not been used.

Soil and water conservation measures could be put into place in hilly areas like the lesser Himalayas, western ghats, etc., where perennial springs and streams exist. In Uttarakhand, 85 % of agricultural lands in high districts are rainfed as the available spring water is not fully utilized. Due to water scarcity, farmers are only able to harvest two crops per year.Utilizing numerous inexpensive poly tanks to collect spring water is suitable for irrigation and domestic uses.

The need of the hour is to formulate rational plan and put them into action quickly. The main issue is that this region could not receive much attention for developmental activities; as a result, the region as a whole still lags in all areas of development. Lack of basic necessities like water, fuel, and health services, rapid population growth and settlement dispersion, small holding sizes, dispersed fields, poor irrigation, a lack of occupational niches, employment opportunities, and work cultures, planning devoid of result-oriented programmes, a lack of local initiatives and investments, the threat of natural hazards like earthquakes, a lack of infrastructural facilities, and lack of topography are the main issues in the area.

The following are the prospects for the state's available resources for sustainable development:

(1) In this environment, horticultural development offers considerable scope and has the potential to become a leadingindustry in many regions of Uttarakhand, notably in the low hills and middle mountains.As indicated previously, various fruits are grown in the area during various crop seasons.

(2) The region is home to a large number of minor forestry products, medicinal plants, and natural fibre plants. These could be used profitably to create valuable and expensive goods.

(3) Promotion of ringal (bamboo) plantation in broader uncultivable regions andusing the material to create useful, marketable products could thus be advantageous from an economic and environmental standpoint. The manufacture of products made from ringals has the potential to develop into a significant industry in particular areas with quite high altitudes. There is plenty of opportunity to expand this business into more lucrative markets, provide local residents with decent jobs, and protect the environment. It

is also known as a plant that ties the earth, reducing landslides and soil erosion.

(4) Hills and mountains at an altitude of 700 m and above are particularly advantageous for raising sheep and goats. The Bhotia households in the high-altitude Pithoragarh, Almora, Chamoli, Uttarkashi, and Dehradun regions have practised it for generations. The State Government's Animal Husbandry Department is working to improve the sheep breeds in the area in order to boost the production of higher-quality wool.

(5) Woolen industries were recently given importance. A larger portion of the population is getting more involved in the manufacturing of woollen goods as a part-time or full-time job, especially women and those with low incomes.

(6) Although tea garden practice has been around for generations, they are far less common now. This practice will help the economic development of the region.

(7) Utilizing water resources to generate hydroelectricity and provide water for cultivation and drinking would pave the road for the area's sustainable development. The introduction of micro hydropower projects is necessary.

Way forward/ Areas of Action

The report identifies three main areas of action for policymakers-

a) Preparing for demographic changes resulting from migrations.

b) We are creating alternative livelihood options in the hill districts to revitalize the economy.

c) I am revisiting the state's agricultural policies.

References

Ahmad, J., Alam, D., & Haseen, M. S. (2011). Impact of climate change on agriculture and food security in India. International Journal of Agriculture, Environment and Biotechnology, 4(2), 129-137.

Arora, N. K. (2019). Impact of climate change on agriculture production and its sustainable solutions. Environmental Sustainability, 2(2), 95-96.

Bhattacharya, A. (2019). Global climate change and its impact on agriculture. Changing climate and resource use efficiency in plants, 1-50.

Dhanai, R., Negi, R. S., Singh, S., & Parmar, M. K. (2014). The effects of climate change on natural resources and socio-economic condition of Himalayan communities of Uttarakhand, India. Int J Mod Commun Technol Res, 2(11), 265749.

Lippert, C., Krimly, T., & Aurbacher, J. (2009). A Ricardian analysis of the impact of climate change on agriculture in Germany. Climatic change, 97(3), 593-610.

Malhi, G. S., Kaur, M., & Kaushik, P. (2021). Impact of climate change on agriculture and its mitigation strategies: A review. Sustainability, 13(3), 1318.

Mall, R. K., Singh, R., Gupta, A., Srinivasan, G., & Rathore, L. S. (2006). Impact of climate change on Indian agriculture: a review. Climatic change, 78(2), 445-478.

Mendelsohn, R. (2009). The impact of climate change on agriculture in developing countries. Journal of Natural Resources Policy Research, 1(1), 5-19.

Mendelsohn, R. (2014). The impact of climate change on agriculture in Asia. Journal of Integrative Agriculture, 13(4), 660-665.

Mishra, D., & Sahu, N. C. (2014). Economic impact of climate change on agriculture sector of coastal Odisha. APCBEE procedia, 10, 241-245.

Molua, E. L. (2007). The economic impact of climate change on agriculture in Cameroon. World Bank Policy Research Working Paper, (4364).

Shakoor, U., Saboor, A., Ali, I., & Mohsin, A. Q. (2011). Impact of climate change on agriculture: empirical evidence from arid region. Pak. J. Agri. Sci, 48(4), 327-333.

Trinh, T. A. (2018). The impact of climate change on agriculture: findings from households in Vietnam. Environmental and resource economics, 71(4), 897-921.

Uleberg, E., Hanssen-Bauer, I., van Oort, B., & Dalmannsdottir, S. (2014). Impact of climate change on agriculture in Northern Norway and potential strategies for adaptation. Climatic Change, 122(1), 27-39.

Yadav, S. S., Hunter, D., Redden, B., Nang, M., Yadava, D. K., & Habibi, A. B. (2015). Impact of climate change on agriculture production, food, and nutritional security. Crop wild relatives and climate change, 1-23.

18

Comparative Economic Analysis of Natural Farming: A Case Study in Una District of Himachal Pradesh

Subhash Sharma[1*], Archita Sharma[2], Chinglembi Laishram[2] and Rohit Vashisth[2]

Dr. YS Parmar University of Horticulture and Forestry, Nauni, Solan Himachal Pradesh, India

Abstract

Natural farming, as an agro-ecologically diverse farming practice, provides numerous ecological and social benefits. In order to better understand the practice followed in NF' as well as the cost saving and income gain by the NF farmers, the study entitled Economic Analysis of Natural Farming: A Case Study in Una District of Himachal Pradesh was carried out during 2019-2020. To calculate its comparative economics, a list of farmers practicing natural farming was taken from the Project Director ATMA, Una, H.P. Thirty farmers from two developmental blocks i.e., Amb and and Gagret were selected randomly. Thus, a sample of 60 farmers was selected for the present study. The results of study revealed the Relative Economic Efficiency (REE) under Natural Farming for different crop combinations was higher as compared to Conventional Farming except Cereal-Vegetables crop combination in Rabi season. The percentage increase in net returns for different crop combinations under Natural Farming over Conventional Farming ranged from 11.93 to 34.41 per cent in Rabi season and 13.42 to 34.64 per cent in Kharif season and also the cost of cultivation for different crop combinations under Natural Farming was less as compared to Conventional Farming system. The per cent reduction in cost of cultivation under NF over Conventional Farming ranged from 9.41 to 17.43 per cent in Rabi Season and 10.86 to 34.27 per cent in Kharif season. This indicated that Natural Farming is low cost farming system as it uses only locally available inputs

Keywords: Natural Farming, Crop combinations, Agro-ecology, SPNF, Himachal.

Introduction

Agriculture is the primary source of livelihood for about 58 per cent of Indian population. Gross Value Added (GVA) by agriculture, forestry and fishing was estimated at Rs. 19.48 lakh crore in FY2020. Growth in GVA in agriculture and allied sectors stood at 4 per cent in FY2020. Share of agriculture and allied sectors in gross value added (GVA) in Indian economy at current prices stood at 17.8 per cent in FY20. Consumer spending in India will return to growth in 2021 post pandemic-led contraction, expanding by as much as 6.6 per cent (Anonymous, 2021). Total food grains production for 2020-21 is set at a record 298.30 million tonnes (MT), comprising 149.92 MT in Kharif season and 148.40 MT during Rabi season. In 2019-20, production was 291.95 MT, against the target of 291.10 MT (Anonymous, 2020b).

As the present form of agriculture is overburdened with inorganic chemicals, fertilizers and pesticides. The presence of toxic chemicals in the atmosphere has been subject of serious debate. Farmers are constantly exposed to high levels of chemicals, typically much higher than customers (Damalas and Koutroubas, 2016). The use of external inputs by adoption of uniform, hybridized and genetically modified crop varieties erodes genetic diversity of seeds and reduces their capacity to adapt to changing climatic conditions. These practices, coupled with wide spread farmland degradation make agriculture a major contributor to global greenhouse gas (GHG) emissions and climate change (Tripathi *et al.* 2018).

So, with such diverse negative impacts of conventional farming, it has become highly essential to adopt other ways which can lead to better results without the negative impact on surroundings. Natural farming is an ideal solution to reduce all these hazards. This sustainable way of farming is also known as 'Do-nothing farming' or 'No-tillage farming'. It was first popularized by Masanobu Fukuoka way back in the 1940s in Japan. It is also known as Zero Budget Natural Farming (ZBNF) in India. It is a method of chemical-free agriculture drawing from traditional Indian practices. It was promoted by Maharashtrian agriculturist and Padma Shri recipient Subhash Palekar, who developed it in the mid-1990s as an alternative to the Green Revolution's methods driven by chemical fertilizers and pesticides and intensive irrigation. The rising costs of these external inputs were a leading cause of indebtedness and suicide among farmers, while the impact of chemicals on the environment and on long-term fertility was devastating. Without the need to spend money on these inputs, the cost of production could be reduced and farming made into a "zero budget" exercise, breaking the debt cycle for many small farmers (Jebaraj, 2019).

Natural farming which is based on agro-ecological principles which emphasizes on the enhanced biomass recycling; enhanced soil-conditions by managing organic matter and soil biological activity and enhanced beneficial biological interactions.

Therefore, NITI Aayog recommended the strong push to be given to ZBNF techniques to reduce costs, improve land quality and increase farmers' incomes (Kumar *et al.* 2019). Looking after ZBNF promotion in India, the Paramparagat Krishi Vikas Yojana, a sub-component of Soil Health Management scheme under the National Mission of Sustainable Agriculture (NMSA), has an outlay of only Rs 325 crore. Out of this, Rs 120 crore has been allocated to the Department of Agricultural Research and Education to conduct experiments at multiple locations and gather findings to establish the viability of SPNF. States such as Andhra Pradesh, Himachal Pradesh, Gujarat, Haryana, Karnataka and Kerala are promoting natural farming. Andhra Pradesh is the frontrunner among all the states in implementing natural farming programme at a mass scale. Himachal Pradesh, too, has set an ambitious target of converting entire state to natural farming by the year 2022.

In Himachal Pradesh, Agriculture and allied sectors contributed to 10 per cent of the total Gross State Domestic Product (GDSP) and provide direct employment to 69 per cent of the population (Anonymous, 2020a). People of Himachal Pradesh are majorly dependent on subsistence farming which is practiced on some irrigated and marginal rain fed farmlands. Himachal Pradesh has adopted ZBNF system in the year 2016-17 and later Himachal Government changed the name to Subhash Palekar Natural Farming. In a perspective to motivate the farmers towards natural farming, the Government of Himachal Pradesh has decided to implement this innovative technique of Subhash Palekar Natural Farming (SPNF) under a new scheme called "Prakritik Kheti Khushal Kissan" to reduce the cost of cultivation and to enhance farm income for the comprehensive and long term welfare and prosperity of farmers and to safeguard them from the adverse climate effects in the year 2018–2019. Cultivation under ZBNF in Himachal Pradesh was started a few years back. Therefore, there is no authentic data on the cost of production/cultivation of various crops grown under natural farming. So, the study was carried out with an objective to compare the economics of Natural and Conventional Farming.

Research Methodology

Una district of Himachal Pradesh was purposively selected for the present study. Simple Random Sampling technique was adopted to select the farmers practising Natural Farming.

A list of farmers practicing natural farming was taken from the Project Director ATMA, Una Himachal Pradesh. Thirty farmers from two developmental blocks i.e., Amband and Gagret were selected randomly. Thus, a sample of 60 farmers was selected for the present study. To meet the objectives of the study, both primary as well as secondary data were collected. Secondary data were collected in terms of no. of registered farmers practicing Natural Farming from Project Director,

ATMA, Una. During the years 2020-2021, primary data were collected from farmers through well-structured and pre-tested schedule. The data from growers included demographic features such as family size, age, education, economic parameters (land use pattern, livestock, cropping patterns, irrigation status, and income), cost of production, yield, farm produce, quantity of inputs used and their prices, and problems faced by farmers in practicing Natural Farming in the study area.

Analytical Framework

To fulfill the specific objective of the study, based on the nature and extent of availability of data, the following analytical tools and techniques have been employed for the analysis of the data.

a) Costs and returns analysis

CACP cost concepts

- **Cost A_1 includes**

  i) Cost of planting material

  ii) Cost of manures, fertilizers and plant protections

  iii) Cost of hired human labour

  iv) Cost of owned and hired machinery

  v) Irrigation charges

  vi) Depreciation on implements, farm buildings and irrigation structures

  vii) Land revenue

  viii) Interest on working capital

  - **Cost A_2:** Cost A_1 + rent paid for leased in land
  - **Cost B_1:** Cost A_1 + interest on the fixed capital assets excluding land
  - **Cost B_2:** Cost B_1 + rental value of owned land
  - **Cost C_1:** Cost B_1 + imputed value of family labour
  - **Cost C_2:** Cost B_2 + imputed value of family labour
  - **Cost C_3:** Cost C_2 + 10 per cent of cost C_2 on account of managerial function performed by the farmer.

b) Crop Equivalent Yield (CEY)

In Natural Farming system, many types of crops were cultivated in a multiple or mixed cropping. So it was very difficult to compare the economics of multiple crops with single crop. Francis (1986) described crop equivalent yield (CEY) to the sum of Equivalent principal and intercrop yields. The differing yields intercrops were transformed into the equivalent yield of any crop depending on the commodity price. So a comparison was made based on economic returns and crop equivalent yields (CEY) of multiple cropping sequences was calculated by converting the yield of different intercrops/crops into equivalent yield of any one crop based on price of the produce. Mathematically, the CEY is represented as :

$$\text{CEY} = C_y + C1_y * \frac{P_{C1}}{P_C} + C2_y * \frac{P_{C2}}{P_C} + ... + Cn_y * \frac{P_{Cn}}{P_n}$$

Where,

- C_y = Yields of the main crop
- P_C = Price of the main crop
- ($C1_{y,}$ $C2_{y, ...}$ Cn_y) = Yields of inter crop, which are to be converted to equivalent of main crop yield
- (P_{C1}, P_{C2}, P_{C3}, P_{Cn}) = Price of the respective intercrops.

c) Relative Economic Efficiency (REE)

Results and Discussion

Cropping pattern of the Sampled Households in the Study Area

Natural Farming

Cropping pattern is the proportion of the cultivated area under different crops at a time. Planting 3-4 crops on the same piece of land, with legumes as the intercrop, is a technique of Natural Farming. As a result, these crop combinations help to enhance the soil characteristics, creates a balance and incorporate organic nutrients into the soil. It was possible to create a complementary relationship between two or more crops when they were grown in close proximity to one another. Number of different crop combinations was examined under Natural Farming and Table1. shows the average area under different crop combinations. Vegetables, Vegetables-Pulses, Cereals-Pulses, and Cereals-Vegetables were the four main crop practiced in the study area. In Rabi season, Vegetables crop combination had a maximum area of 0.11 ha (39.29%) followed by Vegetables-Pulses (35.71%), Cereals-Pulses (17.86 %), and Cereals-Vegetables (7.14%). In case of Kharif season, it was observed that the crop combination Vegetables accounted for highest area

(39.29%) followed by vegetables-pulses (32.14%), cereals-pulses (21.43%) and cereals-vegetables (7.14%). Crop intensification is one of the strategies used to cultivate crops on restricted soil resources which increases yields per unit area and time. In other words, it assures show effectively the land can be used. The cropping intensity was 200 per cent, indicating that the cultivated land is being used efficiently in the study area.

Table 1. Area under major crop combinations under Natural Farming system in the study (ha)

Particulars	Rabi	Kharif
Area	Area	Area
Vegetables	0.11 (39.29)	0.11 (39.29)
Vegetables - Pulses	0.10 (35.71)	0.09 (32.14)
Cereals - Pulses	0.05 (17.86)	0.06 (21.43)
Cereals Vegetables	0.02 (7.14)	0.02 (7.14)
Total Area	0.28	0.28
Gross cropped area	0.56	
Net sown area	0.28	
Cropping intensity (%)	200	

Figures in parentheses are percentage to net sown area

Conventional Farming

In Conventional Farming (CF) sole cropping was practiced while in Natural Farming (NF), three to four crops were cultivated together on the same piece of land. The comparison of CF with NF was carried out by considering one dominant crop of the study area under different crop combinations. The major crops grown under CF system include Maize and Potato in Kharif season and Wheat, Pea and Radish in Rabi season. From the Table 2., it was observed that, in Rabi season, Wheat had contributed a maximum area of (58.90%) followed by Pea (21.92%) and Radish (19.18%). In Kharif season, Potato is the major crop grown under CF system and accounted for 52.05 per cent of the cultivated area under CF system. Maize is the second grown in Kharif season on area around 47.95 per cent of the area under CF system. Cropping intensity under CF system has 200 per cent i.e., area is used efficiently in the study area.

Table 2. Area under major crops under Conventional Farming system in the study area (ha)

Rabi		Kharif	
Particulars	Area	Particulars	Area
Wheat	0.43 (58.90)	Potato	0.38 (52.05)
Pea	0.16 (21.92)	Maize	0.35 (47.95)
Radish	0.14 (19.18)	-	-
Total Area	0.73	0.73	
Gross cropped Area	1.46		
Net sown area	0.73		
Cropping intensity (%)	200		

Season-wise Crop Combinations under Natural Farming System

Multiple crops are grown or cultivated together on the same piece of land under Natural Farming system, along with leguminous crops as intercrops, in order to ensure that no piece of land is wasted and full cultivation of the area. One of the most important components of Natural Farming is intercropping with leguminous crops as it increases soil fertility and productivity through atmospheric nitrogen fixation. Farmers in the study area cultivated different crop combinations SPNF system. The major crop combinations adopted by the selected farmers were i) Vegetables ii) Vegetables-Pulses iii) Cereals-Pulses iv) Cereals-Vegetables.

Table 3. Season wise major crop combinations adopted by selected farmers under NF system

Crop combinations	Rabi season	Kharif season
Vegetables	Pea + Cauliflower	Potato + Pea
	Pea + Garlic	Pea + Coriander + Cabbage
	Pea + Turnip	Potato + Lettuce
	Pea + Cauliflower	Potato + Cabbage
	Potato + French Beans	Potato + Cabbage + Radish
	Pea + Cauliflower	Potato + Coriander
Vegetables - Pulses	Radish + Cowpea	Potato + Cauliflower + Kidney beans
	Radish + Faba Beans	Potato + Soyabean
		Potato + Mustard
Cereals - Pulses	Wheat + Chickpea	Maize + Lobia
	Wheat + Soyabean	Maize + Urad
		Maize+ Soyabean
		Maize + Chickpea
Cereals - Vegetables	Wheat + Garlic Wheat + Turnip Wheat+ Cabbage Wheat + Coriander Wheat + Pea	Maize + Onion Maize + French Beans Maize + Black gram Maize + Cowpea Maize + Green gram

From the Table 3, it was observed that in Rabi season, the major crops being grown in the study area was Wheat, Pea, Radish and other crops includes Cabbage, Cauliflower, Garlic, Turnip, Coriander etc. The main intercrops (leguminous) in the study area included Chickpea, Cowpea, Pea, Faba-Bean and Soyabean. While in Kharif season, Potato and Maize were the major crops and the intercrops included Pea, Lettuce, Cabbage, Radish, Coriander and Beans.

Comparative Analysis of Natural Farming System and Conventional Farming System

Yield

Under Natural Farming system, two or more crops are cultivated together on the same piece of land. NF's economic output was difficult to compare with CF system since multiple crops cultivated in a multiple or mixed cropping system. Crop Equivalent Yield (CEY) was used to compare yield for mixed cropping systems. From the Table 4.2, it was observed that, the yield of all major crop combinations ranged from 31.68 to 43q/ha under Natural Farming system and under Conventional Farming it ranged from 37 to 47 q/ha. The maximum yield was observed in crop combination Vegetables (43 q/ha) in Rabi season. While in Kharif season for all crop combinations, the yield under Natural Farming system ranged from 29.64 to 35.16 q/ha and under Conventional Farming it ranged from 32 to 47 q/ha. The maximum yield was observed in crop combination Vegetables (35.16 q/ha). From the Table 4.2 it was concluded that CEY of the NF system was lesser than CF system. In Rabi season, percentage decrease in NF yield over the CF yield was ranged from 6.82 to 14.38 per cent and in Kharif season it ranged from 2.47 to 29.68 per cent.

Table 4. Crop equivalent yield (CEY) of various crop combinations under Natural Farming and Conventional Farming systems

Crop Combination	CEY (q/ha)					
	Rabi			Kharif		
	NF	CF	Per cent change (NF over CF)	NF	CF	Per cent change (NF over CF)
Vegetables	43.00	47.00	-8.51	35.16	47.00	-25.19
Vegetables + Pulses	41.00	44.00	-6.82	33.05	47.00	-29.68
Cereals + Pulses	33.35	37.00	-9.86	31.21	32.00	-2.47
Cereals + Vegetables	31.68	37.00	-14.38	29.64	32.00	-7.38

Cost of Cultivation

Chemical inputs are a major cost component for cash crops including fruits and vegetables under Conventional Farming system. Because of this farm activity, farmers have to bear higher costs and have witnessed decrease in their income.

Due to the adoption of NF technique, the cost of cultivating these crops has been significantly reduced. Table 5. presented a comparison of cost of cultivation between NF and CF systems. It has been observed that the total cost of all crop combinations in NF systems during the cultivation process was substantially reduced. In the Rabi season, the per cent reduction in NF cultivation costs over the CF system ranged from 9.41 to 17.43 per cent, while in the Kharif season it was ranged from 10.86 to 34.27 per cent. In Rabi season, maximum reduction in cost was observed in crop combination Vegetables (17.43%) followed by Cereals-Pulses (16.90%), Cereals-Vegetables (15.44%) and Vegetables-Pulses (9.41%). whereas in Kharif season, the maximum reduction was observed in crop combination Vegetables-Pulses (34.27%) followed by Vegetables (32.05%), Cereals-Vegetables (12.72%) and Cereals-Pulses (10.86%). Cost of cultivation under Natural Farming was less as compared to Conventional Farming due to use of locally available non-synthetic inputs, which decreases the costs to the farmers. Similar findings were recorded that, the cost of fertilizers was remarkably lesser (almost 90%) in ZBNF than others as they have to buy few natural inputs. So, overall expenditure came out to be lesser for ZBNF farmers than non-ZBNF (Gupta *et al.* 2020).

Table 5. Cost of cultivation of various crop combinations under Natural Farming and Conventional Farming systems

Crop Combination	Cost of Cultivation (Rs/ha)					
	Rabi			Kharif		
	Total Cost		Per cent change (NF over CF)	Total Cost		Per cent change (NF over CF)
	NF	CF		NF	CF	
Vegetables	58,233	70,524	-17.43	44,991	66,215	-32.05
Vegetables + Pulses	52,513	57,971	-9.41	43,520	66,215	-34.27
Cereals + Pulses	51,068	61,456	-16.90	50,383	56,519	-10.86
Cereals + Vegetables	51,969	61,456	-15.44	49,329	56,519	-12.72

Gross Returns

Table 6. shows that the gross returns under NF systems were lesser in all crop combinations compared to the CF system. In the Rabi season, the reduction in gross returns in NF over CF ranged from 6.82 per cent to 14.38 per cent, while in the Kharif season it was ranged from 5.10 per cent to 29.68 per cent. The maximum reduction in gross returns was observed in crop combination Cereal-Vegetables in Rabi season whereas in Kharif the maximum reduction in gross returns was observed in crop combination Vegetable-Pulses.

Table 6. Crop combination wise Gross Returns under Natural Farming and Conventional Farming systems

Crop Combination	Gross returns (Rs/ha)					
	Rabi			Kharif		
	Gross returns		Per cent change (NF over CF)	Gross returns		Per cent change (NF over CF)
	NF	CF		NF	CF	
Vegetables	86,000	94,000	-8.51	52740	70,500	-25.19
Vegetables + Pulses	61,500	66,000	-6.82	49575	70,500	-29.68
Cereals + Pulses	64,198	71,225	-9.86	56178	59,200	-5.10
Cereals + Vegetables	60,984	71,225	-14.38	53352	59,200	-9.88

Net Returns

Thenetincomeofafarmreflectsitsprofitsandlosses.Inotherwords,itisaenterprise's net income after deducting all expenses

Table 7. Crop combination wise Net Returns under Natural Farming and Conventional Farming systems

Crop Combination	Net returns (Rs/ha)					
	Rabi			Kharif		
	Net returns		Relative Economic Efficiency (%)	Net returns		Relative Economic Efficiency (%)
	NF	CF		NF	CF	
Vegetables	27,766	23,475	18.28	7,748	4,284	34.64
Vegetables + Pulses	8,986	8,028	11.93	6,054	4,284	17.70
Cereals + Pulses	13,130	9,768	34.41	5,794	2,680	31.14
Cereals + Vegetables	9,014	9,768	-7.72	4,022	2,680	13.42

By using natural inputs such as *jeevamrit, bijamrit, ghanjivamrit*, etc., input costs are greatly reduced. Table 4.23 revealed that the net returns under NF system were higher than CF system across all crop combinations except in Cereals-Vegetables crop combination where net returns were less in NF than CF in Rabi Season. The Relative Economic Efficiency (REE), the comparative measure of economic gain in NF over the CF, in all crop combinations in the Rabi season was ranged from 11.93 to 34.41 per cent higher except 7.72 per cent reduction in Cereals-Vegetables, while in the Kharif season, it ranged from 13.42 to 34.64 per cent higher in all crop combinations. Maximum Relative Economic Efficiency was observed in the Cereals-Pulses crop combination in the Rabi season and in Kharif season, the maximum was observed in the Vegetables crop combination. Expenditure savings, owing to local inputs and increased income from intercropping have contributed to higher NF returns. As a result of mixed cropping, the farmer was able to make better use of its land and able to raise its net profit, as well as expand the diversity of crops available at different periods during the growing season. The results were supported by the same study undertaken by Koner and Laha (2020) which stated that the farmers faced a reduction in yield initially along with the reduction in production cost. Gradually, the yield has increased and is almost similar to when it was produced by chemical farming and also an increased in income was also noticed as compared to chemical farming. So, this method of farming can be proved to be a good source income generation.

Conclusion

A brief summary of this research on natural farming has been provided in this book chapter. Natural Farming systems have greater net returns over conventional farming system. It is therefore, suggested that government should promote the SPNF model as a way to double farmer's income in long run. Government should promote or establish specialized markets for marketing of SPNF produce so, that farmers must get premium prices with facility of certification of SPNF produce to encourage the farmers' participation in natural farming. Farmers are not adopting complete SPNF model in the study area and unable to get the full benefits of the SPNF system. So, awareness should be created among the farmers regarding importance of using complete SPNF model.

References

Anonymous. 2021. Agriculture in India: Information about Indian agriculture and its importance. https://www.ibef.org/industry/agriculture-india.aspx (8 September 2021)

Anonymous. 2020a. Subhash Palekar Natural Farming Prakritik Kheti Khushal Kisan Yojana, State Project Implementing Unit (SPIU), Directorate of Agriculture, Shimla-5, Himachal Pradesh. https://spnfhp.nic.in/SPNF/en-IN/index.aspx (2020)

Anonymous. 2020b. Govt sets record grains production target for crop year 2020-21. https://www.financialexpress.com (4:35 AM, April 17th 2020)

Damalas C A and Koutroubas S D. 2016. Farmers' Exposure to Pesticides: Toxicity Types and Ways of Prevention. Toxics 4:1-10.

Francis, C. A. 1986. Distribution and importance of multiple cropping. Agricultural System 25: 238–240. doi: 10.1016/0308-521X(87)90024-2.

Gupta N, Saurabh T and Hem H D. 2020. Can zero budget natural farming save input costs and fertiliser subsidies. New Delhi: Council on Energy, Environment and Water 27p. Jebaraj P. What is zero budget natural farming. https://www.thehindu.com (11:45, JULY 28th, 2019)

Kumar R, Kumar S, Yashawanth and Meena P C. 2019. Natural farming practices in India: its adoption and impact on crop yield and farmers' income. Indian Journal of Agricultural Economics 74:420-43.

Koner N and Laha A. 2020. Economics of zero budget natural farming in Purulia District of West Bengal. Studies in Agricultural Economics 122:22-28.

Tripathi S, Nagbhushan S and Tauseef S. 2018. Zero budget natural farming for the sustainable development goals in Andhra Pradesh, India. New Delhi: Council on Energy, Environment and Water 16p.

19

A Case Study on Comparative Economics of Natural Farming in Kangra District of Himachal Pradesh

Subhash Sharma, Pradeep Puria, Chinglembi Laishram and Rohit Vashisth

Department of Social Sciences, Dr Ys Parmar University of Horticulture and Forestry, Nauni, Solan, Himachal Pradesh, India

Abstract

The current input-intensive agricultural system has rising concerns about its sustainability, due to which alternative agricultural systems have been developed. Zero Budget Natural Farming (ZBNF) also known as Subhash Palekar Natural Farming (SPNF) is one of the most recently focused upon farming strategy that is being employed worldwide. Natural farming is a type of chemical free agriculture, which focuses mainly on traditional Indian practices based on agro-ecology and has been adopted to help farmers in India and other parts of the world. The objective was to reduce the cost of production to almost zero and return to the pre-green revolution farming model, which proposes a 100 per cent elimination of synthetic chemical based inputs and promotes use of locally produced inputs. In this book chapter, recent studies of the comparative economics of natural farming and conventional farming have been highlighted. The analysis showed that Natural Farming system provides relatively high returns per hectare than the conventional farming system. The results revealed that farmers witnessed a drop in per hectare cost of production and profitable yield for their crops as well. More importantly, farmers who adopted the SPNF approach have been able to boost their incomes in comparison to their chemical counterparts.

Keywords: Natural Farming, Crop combinations, Agro-ecology, Sustainability

Introduction

Agriculture is one of the most important sector of the Indian economy Indian agriculture sector accounts for 18 per cent of Gross Domestic Product (GDP) and provides employment to 50 per cent of its workforce India is the world's largest producer of wheat, rice, pulses and spices product and second largest producer of fruits and vegetables in the world The food grains production for the year 2019-2020 (291 95 million tonnes) as compared to the previous year 2018-2019 (285 21 million tonnes) was higher by 6 74 million tonnes (Anonymous, 2020).

Currently, India is Asia's largest pesticide manufacturer, and ranks twelfth in the world for pesticide use (Abhilash and Singh, 2009) Conventional agricultural practices have a major impact on the environment Some of the issues related to agriculture are climate change, soil depletion, deforestation, pollution, irrigation problems and waste Excessive use of fertilizers such as urea, nitrogen and pesticides has affected air, water, and soil quality The acknowledgement that Indian agriculture needs to reduce its dependency on chemical fertilizers and adopt sustainable farming practices to take into account the full costs and impacts of existing production practices Sustainable agriculture aims to help the environment by reducing agricultural runoff, reducing contamination of lakes and rivers, conserving water and maintaining soil fertility naturally by recycling nutrients on the farm.

Natural farming is a solution to reduce all these hazards 'Natural farming' means farming with nature and without chemicals It is also known as subhash palekar natural farming (SPNF) It was promoted by Padam Shri Subhash Palekar, which is a self-developing, self-nourishing and self-sufficient farming So, there is no exploitation made by humans in this system It is pain-free, care-free, loan-free, passion-free farming (Palekar, 2010) His opinions that the rising cost of the external inputs is a leading cause of suicide among farmers Without the need to spend money on these inputs, production costs could be reduced which would help many small farmers break the debt cycle (Anonymous, 2016) In contrast to the chemical-based agriculture, the SPNF is expected to bring about changes in the crop pattern from mono to poly- cropping (Galab, , 2018) The main aim of Natural Farming is to eliminate the use of chemical fertilizers and support the use of different formulation made by dung and urine of indigenous cow The main reason to adopt the Natural Farming is to ensure food security and produce more output with less resources This means that farmers *need not to purchase fertilizers and pesticides in order to ensure the healthy growth of crops* Natural farming is comparatively very low in expenditure and production costs compared with conventional farming They can raise fund for initial investment in agricultural operations at lending institutions with government subsidy (Mariyat, 2018) *Natural Farmng is totally different from organic farming* No external inputs such as 'bio–fertilisers', vermi-compost and expensive 'bio' products etc are used in natural farming Soil only provides 1 5-2 0 per cent of nutrients, with the remaining 98– 98 5 per cent provided by the air, water and solar energy These nitrogen in air is fixed by nitrogen fixing bacteria present in soil Therefore the application of 10 kilograms of cow dung and 10 litres of cow urine per acre is adequate to support crop nutrition and multiplication of soil nutrient (Raghuram, 2019).

In Himachal Pradesh, Agriculture is an important sector of the economy, contributing about 45 per cent to the total income (Anonymous, 2019) Farmers of Himachal Pradesh depend on subsistence farming, which is practiced on some irrigated and marginal rainfed farmlands The average land holding in Himachal Pradesh is about 1 2 hectares Due to sub-division of land holdings and inheritance law, number of marginal and small farmers is increasing in the State The total number of marginal and small farmers is about 85 percent of total farms Due to the improvements in yield, small holders and marginal holders can earn more while simultaneously increasing the amount of food available for their families (Anonymous, 2018a).

There are some inputs that need to be purchased from the market The government of Himachal Pradesh, therefore, changed the name of ZBNF to Subhash Palekar Natural Farming (SPNF) In order to encourage the farmers for natural farming , government of Himachal Pradesh has decided to implement farming technique of Subhash Palekar Natural Farming (SPNF) under a new scheme called "PRAKRITIK KHETI KHUSHHAL KISSAN" reducing the cost of cultivation and increasing farm income for the comprehensive and long-term welfare and prosperity of farmers The main motive of this Yojna is to double farmers' income by 2022 and it paved the way to adopt 'Subhash Palekar Natural Farming (SPNF)' in the State as a mean to meet the objectives of the Yojna The state government will provide subsidies, along with financial assistance, to encourage farmers to pursue natural farming 75 per cent of the subsidies on drums required to make inputs for natural farming, financial assistance of Rs 50,000 to open a store of natural resources in the village The budget of Rs 25 Crore has been allocated to the scheme by the State Government for the period 2018-2019 (Anonymous, 2018b).

To understand its impact, the study was conducted with an objective to compare the economics of Natural Farming vis-á-vis Conventional Farming.

Methodology

- Sampling Procedure

a) **Selection of the study area**

Kangra district of Himachal Pradesh was purposively selected for the study

b) Sampling design and sample size

Simple random sampling design was adopted to select the ultimate sample of the farmers practicing natural farming respondents A list of farmers practicing Subhash Palekar Natural Farming was procured from the Project Director ATMA, Kangra Himachal Pradesh Fifteen farmers each from four blocks Nurpur, Fathehpur, Baijnath and Rait were selected randomly Thus, a sample of 60 farmers was selected for the study

i) **Primary data**

Primary data were collected from the farmers practicing natural farming by survey method using well-structured and pretested schedule The data collected revealed the demographic features (family size, age, and education), land use pattern, livestock, cropping patterns, irrigation status, income pattern, farms produce along with quantity of input use and their prices under natural farming and conventional farming system Problems faced by farmers in various aspects of production and marketing of natural farming

ii) **Secondary data**

Secondary data were collected in terms of numbers of registered farmers practicing natural farming from the Project Director, ATMA, Kangra Himachal Pradesh

c) Crop Equivalent Yield (CEY)

In Natural Farming system, many types of crops were cultivated in a multiple or mixed cropping So it was very difficult to compare the economics of multiple crops with single crop Francis (1986) described crop equivalent yield (CEY) to the sum of Equivalent principal and intercrop yields The differing yields intercrops were transformed into the equivalent yield of any crop depending on the commodity price So a comparison was made based on economic returns and crop equivalent yields (CEY) of multiple cropping sequences was calculated by converting the yield of different intercrops/crops into equivalent yield of any one crop based on price of the produce Mathematically, the CEY is represented as:

$$\text{CEY} = C_y + C1_y * \frac{P_{C1}}{P_C} + C2_y * \frac{P_{C2}}{P_C} + \ldots + Cn_y * \frac{P_{Cn}}{P_n}$$

Where,

- C_y = Yields of the main crop
- P_C = Price of the main crop
- ($C1_{y,}$ $C2_{y,} \ldots$ Cn_y) = Yields of inter crop, which are to be converted to equivalent of main crop yield
- (P_{C1}, P_{C2}, P_{C3}, P_{Cn}) = Price of the respective intercrops

Results

Cropping Pattern of Sampled Household under Natural Farming System

The pattern of crops in any region heavily depends on altitude, micro-climate, soil, resource availability and management factors The variations in the gross cultivated area under different farming crop combination indicate the degree of agricultural diversification in sampled farms A close analysis of crop pattern further shows the status of the region's agriculture The proportional share of a specific crop combination in the farm's gross cultivated area shows the value the farmer attaches to a specific system

Natural Farming system is a mix farming system in which the leguminous are grown as a intercrop The cropping pattern of sampled farms was analysed and the results have been presented in Table 1 Table shows that net sown area under Natural Farming was 0 45 ha In Rabi season, three major crop combinations were adopted by the sampled household i.e. 1) Cereals-pulses, 2) Cereals-vegetables,3) Vegetables-pulses Area under different crop combinations was 0.19 ha, 0.16 ha, 0 1 ha respectively 51 66 per cent of farmers adopting crop combination Cereal-pulses followed by Cereal-vegetables (28 34%) and Vegetables-pulses (20 %) Maximum crop combination area under Cereal-pulses (42-22%) followed by Cereals-vegetables (35-55 %) and Vegetables-pulses (22-23 %) In Kharif season, four crop combinations were adopted by the sampled household i e, 1) Cereals-pulses, 2) Cereals-vegetables, 3) Vegetables-pulses, 4) Cereals-vegetables-pulses Area under different crop combination was 0.13ha, 0.12 ha, 0.08 ha and 0.07 ha 30 per cent of farmers adopting crop combination Cereals-pulses followed by Vegetables-pulses (28-34%) and Cereals-vegetables (21-66%) Maximum crop combination area under Cereals-pulses 32 50 per cent followed by Cereals-vegetables (30%) and Vegetables-pulses (20%) and Cereals-vegetables-pulses (17 50%) Gross cropped area of the sampled household was 0 85 ha The cropping intensity of Natural Farming was 188 90 per cent

Table 1. Cropping pattern under Natural farming system of the sampled households

Crop combination	Rabi Season		Kharif Season	
	No. of Farmers	Area (ha)	No. of Farmers	Area (ha)
Cereals- pulses	31 (51.66)	0.19 (42.22)	18 (30)	0.13 (32.50)
Cereals-vegetables	17 (28.34)	0.16 (35.55)	13 (21.66)	0.12 (30)
Vegetables –Pulses	12 (20)	0.10 (22.23)	17 (28.34)	0.08 (20)
Cereal-Vegetables -Pulses	-	-	12 (20)	0.07 (17.50)

Crop combination	Rabi Season		Kharif Season	
	No. of Farmers	Area (ha)	No. of Farmers	Area (ha)
Total	60 (100)	0.45 (100)	60 (100)	0.40 (100)
Net sown area(ha)	0.45			
Gross cropped area(ha)	0.85			
Crop intensity (%)	188.90			

Figure in parentheses is percentage to total

Cropping Pattern of Dampled Household under Conventional Farming

The cropping pattern of sampled farmers under conventional farming was analysed and the results have been presented in Table 2. and shows that net sown area under conventional farming was 0.36 ha. In Rabi season, Wheat and Pea were the two major crops cultivated by the sampled household. Area under Wheat and Pea crop was 0.22 ha and 0.14 ha, respectively i.e. 61.22 per cent area under Wheat cultivation and 38.88 per cent area under Pea cultivation. In Kharif season, Maize and Tomato were the two major crops cultivated by the sampled household. Area under Maize and Tomato crops was 0.26 ha and 0.08 ha respectively i.e. 76.48 per cent area under Maize cultivation and 23.52 per cent area under Tomato cultivation. Gross cropped area of the sampled household was 0.70 ha .The cropping intensity of Conventional Farming was 194.45 per cent.

Table 2. Cropping pattern under Conventional farming system of the sampled households

Crop	Season	
	Rabi (ha)	Kharif (ha)
Wheat	0.22 (61.12)	-
Pea	0.14 (38.88)	-
Maize	-	0.26 (76.48)
Tomato	-	0.08 (23.52)
Total	0.36 (100)	0.34 (100)
Net sown area (ha)	0.36	
Gross cropped area (ha)	0.70	
Crop intensity (%)	194.45	

Season-wise Major Crop under Natural Farming System by the Sampled Household

The main principle of the natural farming is to cultivate the crops in mixed cropping system. Farmers cultivate leguminous crops with the main crops. The main reason behind to cultivating leguminous crop with cereal is that leguminous crop has nodules in root which make a symbiotic relationship and enhances the nitrogen fixation in plant leading to increased productivity of the main crops. It also helps to fix the atmospheric nitrogen in plants. The system of different crop combinations of sampled farms is presented in Table 3. Table shows that, in Kharif season, four major crop combinations in their study area were grown with legumes crops like pulses and vegetables. Maize was the major cereal crop in Kharif season. Legumes crop like soyabean, lentil etc were grown as a intercrop with main crops. In Rabi season, three major crop combinations were grown with legumes crops like pulses and vegetables. Wheat was the major cereal crop in Rabi season. Legumes crop like Black Gram, Beans etc were grown as a intercrop with main crops.

Table 3. Season-wise major crop combinations under Natural Farming

Crop combination	Season	
	Rabi	Kharif
Cereals - Pulses	Wheat+ Chickpea+ Mustard Wheat+ Pea+ Mustard	Maize + Soyabean Maize + Black-eyed pea Maize + Kidney Beans Maize+ Lentil
Cereals - Vegetables	Wheat + Pea +Black Gram	Maize+ Cucumber+ French Bean Maize+ Capsicum+ French Beans
Cereal – Vegetables - Pulses	-	Maize+ Kidney Beans + Okra Maize+ Capsicum+ Mash Maize+ Soyabean+ Okra
Vegetables - Pulses	Peas+ Spinach+ Coriander+ Radish+ Black Gram Pea+ Radish+ Cauliflower+ Beans	Capsicum+ French Beans Tomato+ Soyabean Tomato+ Cucumber+ Soyabean Tomato+ French Beans Tomato+ Capsicum+ Kidney Beans

Comparsion Between Natural Farming and Conventional Farming Systems

a) Yield

In Natural Farming system, crops were cultivated in a multiple or mixed cropping systems. So it was very difficult to compare the yield of multiple crops with single crop. Therefore, a comparison was made on the basis of Crop Equivalent Yields (CEY) of multiple crops. Table 4, shows the comparison of the yield between Natural Farming and Conventional Farming system. Table revealed that there was slight increase in yield in all crop combination system of Natural Farming as compare to Conventional Farming. In Rabi Season, Crop Equivalent Yields (CEY) in Cereals-pulses, Cereals-vegetables and Vegetables-pulses was 44.15 qtl, 42.15 qtl and 77.29, qtl respectively. Maximum yield was of Vegetables-pulses followed by Cereals-pulses and Cereals-vegetables. Percentage change in NF over CF varies between 0.33-0.70 per cent .Maximum increase in yield was found in

Vegetables-pulses system which was 0.70 per cent followed by Cereals-pulses (0.67 %) and Vegetables-pulses (0.33 %). Higher increase in vegetables-pulses can be attributed to the high prices of pulses. In Kharif season, the change in yield of NF over CF varies between 0.79-1.76 per cent. Maximum increase in yield was Cereals- pulses which is 1.76 per cent followed by Cereals-vegetables (1.25 %), Cereals-vegetable-pulses (1.22 %) and vegetable-pulses (0.79 %). The reason behind of increase yield in Cereals-pulses is that pulses crops having root nodules near the root zone which make a symbiotic relationship with the cereal crops and help to fix the atmospheric nitrogen resulting increasing cereal crop production. These yield differences depend on crops which are growing in the system, site characteristics, and price of the intercrops. Similar results were reported from Nabakrushna Choudhury Centre for Development Studies Bhubaneswar, Odisha were the yield/ha of (Paddy, Groundnut. Black Gram, Chilly, Maize) increase 600 kg, 635kg,173kg, 2500 kg and 988 kg respectively as compared to conventional farming. (Mishra S. 2018)

Table 4. Crop Equivalent Yield (CEY) comparison between natural and conventional farming

Crop combination	CEY(q/ha)					
	Rabi			Kharif		
	NF	CF	% increase in NF over CF	NF	CF	% increase in NF over CF
Cereals- pulses	44.15	43.86	0.67	29.07	28.56	1.76
Cereals-vegetable	42.15	42.01	0.33	41.89	41.38	1.25
Vegetables-pulses	77.29	76.75	0.7	113.78	112.89	0.79
Cereals-vegetable-pulses	-	-	-	33.09	32.69	1.22

b) Gross Returns

Gross returns under natural farming system and conventional farming system have been presented in Table 5. Table revealed that in Rabi season, gross returns in Cereals-pulses, Cereals-vegetables and Vegetables-pulses were Rs 79,732, Rs 75,781 and Rs 1,93,535, respectively. Maximum gross returns were Vegetables-pulses followed by Cereals- pulses and Cereals-vegetables. In Kharif season, gross return in Cereals-pulses, Cereals-vegetables, Vegetables-pulses and Cereals-vegetable-pulses were Rs 48,998, Rs 71,054, Rs 2,50,295 and Rs 2,48,349, respectively. Maximum gross returns were Vegetables-pulses followed by Cereals-vegetables-pulses, Cereals-vegetable and Cereals - pulses.

It was also concluded that there was slightly increase in returns in natural farming system as compared to conventional Farming systems. In Rabi Season, Percentage increase in NF over CF varies between 0.2-1.86 per cent. The maximum increasing gross returns were Vegetables-pulses which (1.86 %) followed by Cereals -pulses (1.00 %) and Cereals-vegetable (0.2 %). In Kharif season, percentage increase in NF over CF varies between 0.65 to 1.02, the maximum increase was seen in

Cereals- pulses 1.02 per cent followed by Cereals-vegetable (1.01 %), Vegetables-pulses (0.79 %) and Cereals-vegetable-pulses (0.65 %). Similar results were reported from Andhra Pradesh, that 10 per cent increase in gross income of different crops in Natural Farming over Conventional Farming (Bharucha *et al.* 2020).

Table 5. Gross returns comparison between Natural Farming systems and Conventional Farming systems

Crop combination	Gross Returns (Rs/ha)					
	Rabi			Kharif		
	NF	CF	% increase in NF over CF	NF	CF	% increase in NF over CF
Cereals-pulses	79,732	78,941	1.00	51,993	48,500	1.02
Cereals-vegetable	75,781	75,628	0.2	76,054	70,339	1.01
Vegetables-pulses	1,93,535	1,91,885	1.86	2,50,295	2,48,349	0.79
Cereals-vegetable-pulses	-	-	-	56,258	55,900	0.65

c) Cost of Cultivation

In conventional farming, high prices of pesticides and fertilizers result in increased the input cost. But in natural farming, all inputs are prepared at home by using domestic raw material like cow dung, urine, jaggery etc so the cost of cultivation under natural farming is very low as compared to conventional farming system. Table 6. shows the total cost of cultivation under Natural Farming and Conventional Farming system. Table revealed that in Rabi season, total cost in Cereals-pulses, Cereals-vegetables and Vegetables-pulses were Rs 49,695, Rs 46,227 and Rs 55,828, respectively. Maximum cost was found in Vegetables-pulses followed by Cereals-pulses and Cereals-vegetables. In Kharif season, total cost in Cereals-pulses, Cereals-vegetables, Vegetables-pulses and Cereals-vegetable-pulses was Rs 45,964, Rs 53,859, Rs 55,939 and Rs 48,766, respectively. Maximum cost of cultivation was found in Vegetables-pulses followed by Cereals-vegetable, Cereals-vegetables-pulses, and Cereals-pulses.

Table 6. Comparison of cost of cultivation of natural farming and conventional farming system

Crop combination	Total Cost of cultivation (Rs/ha)					
	Rabi			Kharif		
	NF	CF	% increase in NF over CF	NF	CF	% increase in NF over CF
Cereals- pulses	49,695	53,577	-7.2	45,964	43,569	-4.02
Cereals-vegetable	46,227	51,470	-13.71	53,859	55,430	-2.83
Vegetables-pulses	55,828	78,850	-29.19	55,939	59,780	-6.42
Cereals-vegetable-pulses	-	-	-	48,766	49,800	-2.07

Table 6. showed that in Rabi season, percentage decrease in cost of cultivation in NF over CF varied between 7.2 to 29.19 per cent.The maximum reduction was seen in Vegetables-Pulses (29.19 %) followed by Cereals-vegetables (13.71 %) and Cereals-pulses (7.2 %). In Kharif season, percentage decrease in cost of cultivation in NF over CF varied between 2 to 6 per cent. The maximum reduction was seen in Vegetables-pulses (6.42 %) followed by Cereals-pulses (4.02 %), Cereals-vegetable (2.83 %) and Cereals-vegetable-pulses (2.07 %). It was concluded from the table that the cost of cultivation of different crop combination under NF was lower as compared to the CF systems. Similar results were found in Karnataka, the cost of cultivation of different crops were lower in Natural Farming as compared to Conventional farming. The cost of (Paddy,Guli Ragi, Ragi and Black Gram) i.e Rs. 30,983 , Rs. 7,375 , Rs 6,875 and Rs 15,775 in natural farming and Rs 43,839, Rs 8,125, Rs 7,625 and Rs 18,595 in Conventional Farming. (Khadse *et al.* 2019).

d) Net Returns

Net returns are the difference between Gross returns and Cost of cultivation. In Natural Farming, input cost is very low as compare to conventional farming. There is slightly increase in gross returns of natural farming as compare to conventional farming system. Lower input cost in natural farming leads to increased net returns as compared to conventional farming. Table 7. shows the net returns under natural farming and conventional farming system. Table revealed that in Rabi season, net returns in Cereals-pulses, Cereals-vegetables and Vegetables-pulses were Rs 30,037, Rs 29,554 and Rs 1,37,707, respectively. Maximum net returns were found in Vegetables-pulses followed by Cereals-pulses and Cereals-vegetables. In Kharif season, Net returns in Cereals-pulses, Cereals-vegetables, Vegetables-pulses and Cereals- Vegetable-pulses was Rs 6,034, Rs 22,195, Rs 1,94,356 and Rs 7,491, respectively. Maximum net returns were found in Vegetables-pulses followed by Cereal-vegetable, Cereal-vegetables-pulses, and Cereal-pulses.

Table 7. conclude that in Rabi season, Relative Economic Efficiency (REE) varied between 18 to 34 per cent. In various crop combinations, the maximum increase in net returns was of Cereal- vegetables (34.02 %) followed by Vegetable-pulses (21.82 %) and Cereal-pulses (18.42 %). In Kharif season, Relative Economic Efficiency (REE) varied between 3 to 28 per cent. The maximum Relative Economic Efficiency (REE) of Cereals-pulses (28.41 %) followed by Cereals-vegetable-pulses (24.24 %), Cereals-vegetable (12.26 %) and Vegetables-pulses (3.06 %). These results were supported by a study in Karnataka, where the net returns of different crops were high in natural farming as compare to conventional farming. The net returns of (Paddy,Guli Ragi, Ragi and blackgram) i.e Rs.60,743 , Rs. 42,789, Rs 31,590 and Rs 39,034 in natural farming and Rs 40,335 ,Rs 27,717, Rs 25,195 and Rs 27,243 in conventional farming. (Khadse A. et al 2019).

Table 7. Net returns comparison between natural and conventional farming systems

Crop combination	Net Returns (Rs/ha)					
	Rabi			Kharif		
	NF	CF	Relative economic efficiency (REE)	NF	CF	Relative economic efficiency (REE)
Cereals- pulses	30,037	25,364	18.42	6,034	4,699	28.41
Cereals-vegetable	29,554	22,051	34.02	22,195	19,770	12.26
Vegetables-pulses	1,37,707	1,13,035	21.82	1,94,356	1,88,568	3.06
Cereals-vegetable-pulses	-	-	-	7,492	6,029	24.24

Conclusion

In this book chapter, we provided a concise note on the research on natural farming. Intercropping with leguminous crop is considered to be one of the most important components since it increases its productivity and fertility of the soil by nitrogen fixation. The research focused on many of the problems that have arisen in the development of natural farming system and the marketing of natural produce in the study area. The findings of this study have led to a wide range of policy consequences for natural farming systems. On the background of the study, it was concluded that cultivation by natural farming method has been found to be productive in the study area. However, in the case of natural farming system the farmers used their own resources in the form of biological inputs, crop preservation and high use of family labour. Farmers following these methods have high yield and returns. This indicated that there is still possibility of rise in income from the adoption of natural farming through the use of own capital and a stronger marketing method. It is therefore suggested that a liberal policy be established to provide guidance, in order to increase the adoption of natural farming technologies in the rural area.

References

Abhilash P C and Singh N. 2009. Pesticide Use and Application: An Indian Scenario. Journal of Hazardous Materials 165: 1-12.

Anonymous, 2019. Economic Survey of Himachal Pradesh 2018-2019. https://himachalservices.nic.in/economics/en-IN/economic-survey.html (7:30 PM, 15th July 2020)

Anonymous. 2016. Zero Budget Natural Farming in India. FAO, Rome. 4p.

Anonymous. 2018a. Information and public relations Himachal Pradesh.http://himachalpr. gov.in/PressReleaseByYear.aspx?Language=1&ID=11892&Type=2&Date=25/04/2018(4:30 PM, 10th July 2020)

Anonymous. 2018b. Prakritik Kheti Khushhal Kissan in Himachal Pradesh 2018. Department of Agriculture, Himachal Pradesh. 12p.

Anonymous. 2020. India's 2019-20 foodgrain production to hit a record high of 291.95 million tonnes. https://economictimes.indiatimes.com/news/economy/agriculture/indias-2019-20-foodgrain-production-to-hit-a-record-high-of-291-95-million-tonnes-estimates-second-advance-estimate-of-govt/articleshow/74192668.cms?from=mdr (8:30 AM, 17th July 2020)

Bharucha Z P, Mitjans S B and Pretty J. 2020. Towards redesign at scale through zero budget natural farming in Andhra Pradesh, India. International Journal of Agricultural Sustainability18: 1-20.

Francis, C. A. 1986. Distribution and importance of multiple cropping. Agricultural System 25: 238–240. doi: 10.1016/0308-521X(87)90024-2.

Galab S, Reddy P P, Raju S R, Ravi C and Rajani A. 2018. Impact Assessment of Zero Budget Natural Farming in Andhra Pradesh – Kharif 2018-19, Centre for Economic and Social Studies, Nizamiah Observatory Campus, Begumpet, Telangana. 60p.

Khadse A and Rosset P M. 2019. Zero Budget Natural Farming in India-from inception to institutionalization. Agroecology and Sustainable Food Systems 8: 2168-3573.

Mariyat P. 2018. Zero Budget Natural Farming (ZBNF) Boon or Bane - A Study with Reference to the Agriculture Sector. Journal of the Gujarat Research Society 21: 69-73.

Mishra S. 2018. Zero Budget Natural Farming: Are This and Similar Practices The Answers, Nabakrushna Choudhury Centre for Development Studies, Bhubaneswar, Odisha. 23p.

Palekar S. 2010 Zero Budget Spiritual Farming. http:// palekarzerobudgetspiritualfarming.org/ home.aspx (11:00 AM, 5th July 2020)

Raghuram N. 2019. Zeroing in on farm budgets or zero budget natural farming? A perspective from India. Perspectives (UNEP). 9p.

20

Agripreneurs and Speciality Cultivation: The Valuable Insights

Shelleka Gupta, Bonia Sharma and Ronnie Dutt

Abstract

Entrepreneurship and innovation are the driving fuel of economic growth. An entrepreneur literally means an individual who creates and invests in one or more businesses, bearing most of the risks and enjoying most of the rewards. Welcoming the concept of entrepreneurship in the agriculture is being more fruitful as agricultural entrepreneurship is the basis of creativity, innovation, profitability and risk management of agricultural producers which leads to survival of population. The term has been given the name of AGRIPRENEUR. Although the agricultural entrepreneurship is less used in Indian scenario, agriculture being old primary activity. But due to the onset of concept of speciality agriculture, the old age prejudice about the agriculture is getting vanished slowly. Moreover, speciality agriculture has embraced the entrepreneurial development in more comprehensive manner. Thus, this paper will emphasize on entrepreneurship development in speciality agriculture due to the fact that the contribution of speciality agriculture has shown growth potential in gross domestic product of the nation. Also, the study will put light on the few case studies of the speciality agriculture with respect to Jammu region of UT of JK. The aim of the paper is to look into the various facets of the speciality agriculture for the agripreneurial growth.

Keywords: *Entrepreneurial Development, Agripreneurs, Speciality Agriculture*

Introduction

Being a global agricultural powerhouse, agriculture sector's importance in India's economy goes well beyond imagination. As, our nation have second-largest agricultural land in the world, with around 60% rural Indian households making their living from agriculture. In India, the agriculture sector employs half of country's growing population and rest of the population is greatly dependent on the farmers for subsistence. It has been the primary sector of India serving people from times immemorial till now.

Budget experts of India have estimated the agricultural growth rate to be near to 3.9 per cent in FY 2022-23, surpassing the 2021-22 growth rate of 3.6 percent approximately. The government of India's priority shall continue on doubling farmers' income by 2022-23 and becoming a USD 5 trillion economy by 2024-25.According to economic survey 2022, the share of the sector in total gross value added (GVA) of the economy has a long-term trend of nearly 18 percent. According to the estimate, the share of the agriculture and allied sector in total GVA has improved to 20.2 percent in 2020-21 and 18.8 percent in 2021-22. Thus, Indian agriculture sector has experienced buoyant growth in the past two years. Moreover, there has been marked shift in production from basic food grains to high value produce which embrace the special agriculture.

Speciality agriculture consist of commercial agriculture operations, rather than a confined feeding operations, which requires smaller tracts of land due to the intensive nature of the operations. Specialty crops defined by US department of agriculture as"fruits and vegetables, tree nuts, dried fruits and horticulture and nursery crops, including floriculture." These may include variety of crops like, coffee, cut flowers, dry edible beans, dry peas, mushrooms, peppermint, sweet corn, lavender, lentils, herbs, honey etc. The specialty crop sector has seen and been consistently growing over the last decade at national and international level. Looking into growth perspective of the speciality agriculture, new entrepreneurial activities seems to make place in the market. As, it would benefit the entrepreneur on individual scale and also increasing the production of these specialty crops would contribute to economy. These additional benefits would increase economic activity via new entrepreneurial development.When the productivity of global agriculture increased drastically as a result of new advance, it is believed to be due to entrepreneurs who are the key drivers of innovation and integral to creating a thriving economy (Uplaonkar&Biradar, 2015).

In the last decade, agriculture sector has witnessed the inflow of educated entrepreneurial youth with ideas, innovations for making speciality agriculture successful venture. Here, the contemporary term named AGRIPRENEUR shows importance. It the combination of two words, i.e. Agriculture + Entreprenures = Agripreneur.Agripreneurship is the entrepreneurial process taken up in agriculture and is the process of adopting new methods, processes, techniques in agriculture or the allied sectors of agriculture for better output and economic earnings. Agripreneurship converts agricultural activity into an entrepreneurial activity (KK Chand 2020).In this journey, start-ups are proving the bridge between the valuable ideas of the agripreneurs and desired products to consumers on the other hand.

The start-ups are examples of how the great things are done by a series of small things brought together. By taking small steps, moving from one phase to another and solving the issues by innovation is what these entrepreneurs are trying to do in speciality agriculture.

Therefore, in the present conceptual paper, the focus will remain towards the agripreneurial growth with speciality agriculture. Moreover, the aim of paper is to put light onvarious facets of the speciality agriculture for agripreneurs to bloom and prosper in India.

Background

Various scholars have given the meaning and definitions of agricultural entrepreneur i.e. agripreneurs in their own way accordingly. In this context, Dollinger (2003) have defined that entrepreneurship in agriculture is the creation of innovative economic organization for the purpose of growth in agriculture keeping in mind the risk associated. Also, Sudharani (2002) stated that agripreneurship as "generally sustainable, community oriented, directly marketed agriculture. Sustainable agriculture denotes a holistic, systems-oriented approach to farming that focuses on the interrelationships of social, economic and environmental process."

As our country is witnessing growing unemployment and poverty and moreover, slow growth of agriculture, there is need of entrepreneurship in agriculture for more productivity and profitability of agriculture with innovative new ideas. The agripreneurship, seems to be essential to develop entrepreneurs and management workforce to cater agricultural sector's success (Kular& Brar, 2011).Agriculture and allied sectors are considered to be mainstay of the economy of India because these are important sources of raw materials for industries and they demand for many industrial products particularly fertilizers, pesticides, agriculture implements and a variety of consumer goods (Bairwa et al., 2014).Agripreneurs are looking specialty farming as boon for their success in agriculture sector as specialty crops have been gaining attention across the agricultural spectrum. The statutory definition of the term "specialty crop" is wide-ranging, including fruits, vegetables, tree nuts, dried fruits and vegetables, horticulture, etc.

Agripreneurship would foster the growth and development of specialty farming and ultimately national economy through entrepreneurship development which increases the income level and employment opportunities in rural as well as urban areas. They wouldhelp in inducing productivity gains by other small farmers who preferred speciality crops and integrating them into local, national and international markets. Through their new and unique innovative ideas agrepreneurs would help in reducing food costs, supply uncertainties and improving the diets of the rural and urban poor in the country.

India has been considered an agrarian economy till today (Misra&Puri, 2005). Opportunities in agriculture and allied sectors can be identified at different phases of agricultural process. Basically, the agripreneurial opportunities are at every stage of agriculture like input stage, farming stage, value chain, output processing and marketing stage and related services. Presently, the scope and potential of agripreneurial opportunities is increasing as a result of globalization and more interconnected world market at global platform. Agriculture process needs many kinds of inputs likeseeds, fertilizers, pesticides and innovative and localized farm technology. So,these areas create opportunities for agripreneurs for developing and producing these inputs. There are very promising opportunities for entrepreneurial process like innovation in the way of using new bio-pesticides, bio-fertilizers,vermicomposting, testing and a mending soil etc. Moreover, the increasing focus on organic farming is opening still more opportunities. There is a lot of scopefor research and development with respect to seed development. Innovative skills of agripreneur works here in developing varieties of seeds that are expected to serve even in unfavourable climatic conditions. From the seeds to the market place, agripreneur tends to show his/her effectice traits in speciality farming. For realizing maximum revenue and improving living conditions of our farming community, productivity of the crops should be improved which is possible with good management practices alongwith good quality of inputs.

Further, there is a lot of entrepreneurial potentialin the area ofagro- tech products. There is a gradual shift happening from the usage of chemical intensive fertilizers and pesticides to natural manure and pesticides. This gradual shift is again opening huge potential and opportunities for agripreneursfor production and marketing of bio-pesticides, eco-friendly agrochemicals and natural manures. At the stage of farming, the chief objective is to maximize the output and leveraging the advantage of seasons. There is an opportunity to go innovative about balanced use of fertilizers and pesticides, agrochemicals, adopting multiple cropping and crop rotation to protect and preserves and enhance the quality of soil, using the agri-tech machines to reduce the cost of cultivation and of labour etc. Farming stage is the actual place to show one's potential and effectiveness as an agripreneur, as it is working platform.

In the post- harvest, the last stage of cultivation business, the opportunities are felt in the areas of value chain, output processing and marketing. The changes happening in supply chain management of agriculture produce are opening doors for new businesses. Agriproducts processing units are booming up. There are opportunities in the area of distribution and logistics (Pandey, 2013). To specify, for agripreneurship development, there are opportunities in the areas of farming vegetables, fruits, food grains, pulses, oil seeds etc., developing greenhouse concept, herbal plantation, dairy and poultry development, animal husbandry, grading and packaging of agri-products, establishing food processing units, establishing cold. From, the above discussion it can be concluded that if each stage has challenges of its own then it is accompanied by the opportunities as well which an efficient agripreneur needs to grasp for success of its farming.

Need/ Scope of Agripreneurship

Entrepreneurship is indeed very crucial for the survival and growth of the cultivators in a changing economic environment (McElwee 2006; Wolf &Schoorlemmer 2007). So, the agripreneurial activities and the importance of the management skills, human relationships, and new forms of marketing developed by the agripreneurs for the economic sustainability of the farm are key to make small farms competitive (Kahan 2013; Migliorini et al. 2018). Agripreneur would make every possible effort to make specialty agriculture and its products locally available that too at lowest cost.Agripreneurs are taking the special crops individually and are trying to make these crops market as niche market which proves to be very competitive and hard for competitors to match. These crops when become a place or location specific with differentiated features, get GI (Geographical Indication) by govt. such as JK UT's Saffron, Nagpur' Oranges, Assam Tea etc. Further, it is seen that all 15 major climates of the world, from snow bound Himalayas to hot humid southern peninsula, from Thar Desert to heavy rain areas all exist in India. There are 20 agro climatic regions and nearly 46 out of 60 soil types of world in our nation. India is the centre for biodiversity in plants, animals, insects, micro-organisms and accounts for 17 per cent animal, 12 per cent plants and 10 per cent fish genetic resources of the globe. In recent years, there has been a considerable emphasis on crop diversification or speciality farming i.e. towards horticulture (fruits, vegetables, ornamental crops, medicinal and aromatic plants and spices), plantation crops (coconut, cashew nuts and cocoa) and allied activities (Tripathi& Agarwal 2015).

Also, these agripreneur would find easy ways through technology and innovation that would not require huge infrastructure that are economically viable and ecologically sustainable too in a country like India. Moreover, agripreneurship development has huge potential of creating new employment opportunities for rural youth. Furthermore, agripreneurship helps in checking migration of rural youth from villages to urban centers and helps in improving living condition of farmers by providing alternative source of income.

Due to the hard efforts of the government of India, cultivable land area under speciality farming has increased from 12.27 lakh hectare in 2018 to 29.39 lakh hectare in 2022. Over the years, the organic promotion of specialty farm activities led to development of state specific organic brands, increased domestic supply. So, from the success of the organic initiatives, a target of 20 lakh hectare additional area coverage by 2024 is envisaged in the vision document. The Government of India provides assistance for promoting specialty and organic farming across the country though different schemes.

Moreover, the Government of India has launcheda Central Sector Scheme of "Formation and Promotion of 10,000 "Farmer Producer Organizations (FPOs)" to form and promote 10,000 new FPOs till 2027-28.Under the scheme, the formation and promotion of FPO is based on Produce Cluster Area approach and specialized commodity-based approach. While adopting cluster based approach, formation of FPOs focuses on "One District One Product" to enable product specialization. Also, the Ministry of Food Processing Industries provides financial assistance in the form ofgrant-in-aid. Other agencies of the Government like National Horticulture Mission (NHM), National Horticulture Board (NHB), Agricultural and Processed Food Products Export Development Authority (APEDA), National Cooperative Development Corporation (NCDC) so on and State Government's also provide assistance for cold storages under their respective schemes. Therefore, some of the schemes of Government of India in supporting the agriculture business sector, specifically in fruits and vegetables preservation and storage and handling which willdefinitely boost the morale of the farmers and other stakeholders to grow in theagri-business and establish agripreneurship which proves to be rising area of economic growth.

Success factors ofAgripreneurs

Entrepreneurship in the agricultural sector is among the best solutions that can both help the growth and development of the agricultural sector and significantly reduce environmental crises (Saridakis et al., 2021).Agripreneurs must possess some traits that could fetch the glory to their work i.e. specialty farming or agriculture. Agripreneur in speciality farming is visionary in himself who can see the particular special cultivation as his leading agribusiness. Also, agripreneur as a visionary set example for other cultivators to move on his footsteps. Risk taking potential is the foremost essential quality or success factor to be an agripreneur as the agrarian sector of India is full of unpredictability. In connection with this, optimism must come hand in hand, due to unpredictablity of success or failure of speciality farming where agripreneur must not lose hope instead should see brightness of success coming. Moreover, agripreneurs must have genes of the competitiveness that would see growth of agro-processing enterprises by advancing innovation in products, processes. These agricultural entrepreneurs should be capable enough to see the challenges and barriers in the specialty farming and further convert them into the success factors itself.It is seen that agriproducts processing units are booming up due to high morale of speciality agripreneurs (Pandey, 2013). Working hard is very cruical not only in agriculturebut in every field of life. But in speciality farming, an agripreneur needs to put even more efforts to raise the productivity. Therefore, it can be seen that successfactors are the essential criteria to be an agripreneur in speciality farming.

To specify, for agripreneurship development, there are opportunities in the areas of farming vegetables, fruits, food grains, pulses, oil seeds etc., developing greenhouse concept, herbal plantation, dairy and poultry development, animal husbandry, grading and packaging of agri-products, establishing food processing units, (Alex 2011). The agripreneur here must be opportunity seeking which could then be converted into success agri-business. Broadmindedness of the agripreneur in speciality agriculture also serve as a success factor for him as if one way of doing things doesn't work, he must fetch another innovative way to accomplish task. So,it can be seen that specialty farming by agripreneur do need various success factors discussed above.

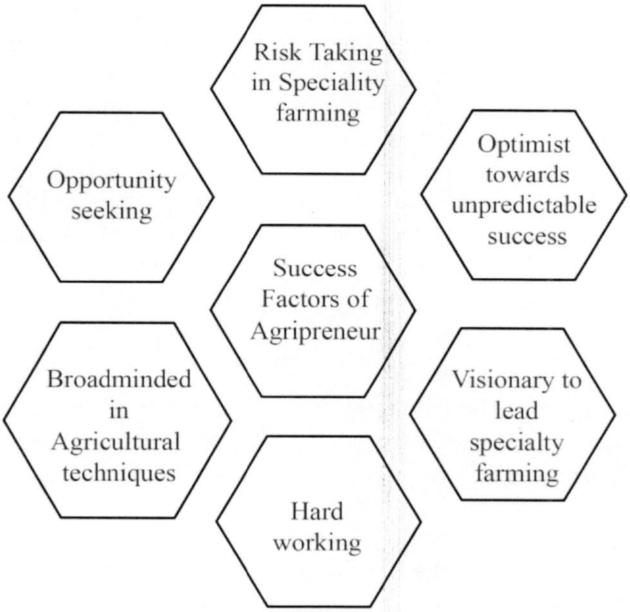

Challenges for Agripreneurs

Agripreneurs must be strong enough to face the challenges coming into the way of specialty farming. Specialty crops include many vegetable and fruit plants, each with their own environmental ranges. The most important and foremost challenge for special crop cultivators is the location of crops i.e. the natural habitat of the specialty species of crops. Climate change will have an impact on these special crops. The major climatic factors are temperature and water in context of Indian Agro farming (Hatfield &Walthall, 2014). Temperature is a major environmental change expected to affect production of annual specialty crops as temperature responses vary more widely among specialty than other crops. Also, water is a critical component in specialty crop production and maintaining an adequate water supply is necessary to achieve high yields and product quality acceptable for consumers (Hatfield &Walthall, 2014).

Furthermore, these people would face challenges like lack of financial resources, incompetent labour force, legal formalities, technological refrains, lack of entrepreneurial culture among people,migration of skilled and talented workforce from rural area to urbanetc. But all these are acceptable and can be setback by govt., collective or individual efforts. These challenges should not be allowed to become huge problem in the success path of speciality farming rather these must be overcome to reach the maximum growth in the specific field. However, the COVID-19 pandemic has presented unique challenges to the industry that threaten its continued growth, while at the same time prompting a positive trend in consumer preferences for direct farm specialty crop purchases. Another, challenge for agripreneur is bonding together the production, processing, or marketing cooperatives which if tackled in better way can increase their relative economic power. Performing some processing functions on-farm before selling their crops, producing higher-value specialty crops, building direct marketing opportunities that would takeout middlemen, and looking for niche markets are other challenges that farmers face and moving out of these challenges can help to capture a larger share of the economic value of what they produce.

Suggestions

After deep studying into the various sides of the agripreneurs in specialty agriculture, the study ought to show avenues or provide suggestions to the prospective readers. Firstly, there is need to change the outlook of the common people about the farming. People still think that agriculture is for those who couldnot achieve heights in other sectors. There is need for promoting agripreneurial culture among the people and to create a vibrant environment for the development identifying promising agricultural and allied areas of business to promote entrepreneurial activity. Secondly, the demand of the hour is to provide area specific technical training programmes to develop the required technical competency among the potential agripreneurs. Then, govt. must intervene by promoting the establishment of forum and platform where organization concerned with entrepreneurial development in agriculture. Also, establishing developmental fund to support the start-up process of agripreneurship to establish agricultural incubation centers. Further, extending support in terms of providing financial and marketing support entrepreneurial education and training to the potential rural youth.Moreover, identifying the trust areas for agripreneurial development across the stages of value chain of the agricultural process and designing appropriate strategies for the promotion. Furthermore, improving infrastructural setup in rural areas identification of important and specific areas of agripreneurship development across the stages of value chain of the agricultural process and designing appropriate strategies for the promotion. Also, providing entrepreneurial education and training to the potential rural youth improving the infrastructural facilities in rural areas, so that entrepreneurial activity can be initiated. Lastly, govt. must keep its door open for all kind of support system to the potential agripreneurs.

A shift from agriculture to agribusiness is very important pathway to bring change or makerevolution in Indian agriculture for the economic and social growth. Agripreneurship hasgot power to generate growth, diversifying income, providing widespread employment andentrepreneurial opportunities. Following case studies make it clear that speciality farming may sound less growth potential but together these can bring a huge change.

A Case Study of Marigold Cultivation

Devender Kumar from district Doda of the Jammu region, UT, JK had decided to undertake marigold cultivation at his farm. He got knowledge and technical information from his relative uncle and under took marigold cultivation onhis land. He raised nursery of hybrid yellow marigold at his farmand thereafter planted them and earned 0.49 lac as the netprofit during the first year of his cultivation. He went on increasing area under marigold year afteryear and that two with different varieties and colours. Every year, he invests Rs 0.5 lac in marigoldcultivation for buying quality seeds, land suitability fertilizers etc. He is adopting the recommended package of practices for cultivation of the marigold crop. He oftenly consults the experts and other experienced farmers of the village. Being the agripreneur, in the area of his residence, his growth in the speciality cultivation of marigolds is increasing with every passing year. During COVID-19 crisis, initially he suffered some losses but later, due tonon-transportation of the flowers from outside the state, the rate of marigold inJammu market has risen and marigold farmers of Doda are making huge profits.

He faced many constraints when he decided to undertake marigold cultivation at his farm in addition to crops, horticulture and animal husbandry, his family was doubtful of hissuccess but they supported him and provided them full assistance during transplanting and plucking flowers. Non-availability of quality marigold seed inthe local market, high fluctuations in the market price, non-availability of localmarket, high cost of inputs especially seed and transportation related problemswere reported to be the major constraints encountered by him in his endeavor. But he being agripreneur learnt from techniques of speciality cultivation of marigold before he fully got into its cultivation.

Devender Kumar always urges the other farmers to undertake speciality agriculture of marigold with major emphasis on proper seedrate and line sowing. He looks this cultivation has much potential of profit earning as he shared his views, “I am a small famer and my income before adopting marigold cultivation was quitelow. Marigold crop has helped me to increase my farm income and I recommend small and marginal farmers of District Doda to adopt it. Inspired by him, many farmers in the adjoining villages have also started marigold cultivation.

Moreover, he has planned to take much of his land undermarigold cultivation. Also, local govt. is taking initiative to develop these adjoining villages which had adopted speciality agriculture of marigold, as farmers here face a lot of difficulties in the transportation of the flowers from here to Jammu. Further, the agripreneurs here expect that the government should set up a processing plant inthe district for extraction of natural colour of the marigold.

Hence, Devender Kumar is an agriprenuer who has raised his farm income to Rs 1.8 lac throughadoption of marigold cultivation at his farm. The marigold cultivation in District Doda is gaining momentum day by day. The success of one marigold farmer has been getting replicated to the fields of other small and marginal farmers and providing spinal cord support to the farm families'economies.

A Case Study of Post Harvest Handling of Papaya

For healthier business, the agriculturist needs to have a greater vision of possibilities. This may require a broader awareness of buying, selling and marketing. It includes various factors like methods of cultivation, modes of transportation, level of packing during transport, final condition of product ensuring customersatisfaction. Fruits and vegetables form an important part of speciality agri-business. Most of the small agriculturist lacks behind with improper packing techniques during transportation and handling in the current market leading topost-harvest losses and dissatisfaction among consumers thereby resulting in downfall of business. This issue occurs in soft fruits and vegetables like papaya, banana, mango, guava, citrus fruits,watermelon, tomato, cauliflower, carrot etc., as even a small mistake in handling these products can damage the whole produce. All the fruit and vegetables handled without proper packing system are prone for mechanical and other physical damages, which cause surface and internal damage furtherleads to spoilage and thus loss. Now, here the smart mind of agripreneur works. He doesnot let the hard efforts of the farmers go waste rather he brings up new ideas for the problems.This problem can be resolved by adopting anappropriate packing technique using plastic perforated crates in place of jute bags or dumping the produce in open carts. Various countries have various methods of packing depending on their sustained conditions.

For this case, the details of different ways of handling of papaya and banana were collected by visiting to the local retailers and roadside vendors around Jammu and its outskirts. Current technique of handling and packing the papayas and bananas in jute bags/plastic bags or open carts is found to be uneffective. Even though carrying/packing/handling materials like plastic/jute bags is cheaper, there are problems associated with these i.e.

1. Contamination of the fruits due to unhygienic conditions of the packing material.
2. These are poor packing materials for fruits which cannot provide protection to the fruits and vegetables.
3. Spoilage and quality losses due to bruising, abrasion, localized unwanted gas formation and ripening.
4. More fundamentally, loss of produce cultivated with hardship of the farmer, but remains unconsumed and simply goes to garbage which means loss despite of hard efforts of farmers.

Better Method Suggestions

1. Plastic crates/Fibre board boxes for transportation and handling as these are self-stackable crates which will eliminatethe excessive stack weight, because the load of the fruits will not fall on the bottomfruits.
2. Perforated crates will allow the proper ventilation ofnecessary gases(oxygen, carbon dioxide,ethylene) which reduces contamination due to effective use of plastic crates.
3. Average savings using this method are 30 to 40% recovery of wastage and increase of income to the farmers, vendors and better quality fruits and vegetables to theconsumers.

Therefore, from the above discussed case studies of agripreneur of speciality farming it becomes clear that agrepreneurship is difficult task, one must believe in one's moves as a visionary to be an agripreneur.

Discussion and Conclusion

It is the known fact that agriculture sector has a large potential to contribute to the national income while at the same time providing direct employment and income to the possibly larger and vulnerable section of the society. Agripreneurship is not only an opportunity but also a necessity for improving the production and profitability in agriculture sector. The economy of India will get boost from the agripreneurship with the identified strengths and to achieve tremendous growth in primary sector and to contribute for rural development. The government has to make policies for focused agripreneurship development and establish developmental organizations to support the process. Surprisingly, agripreneurship development is going to give excellent results at rural level and urban level also. In this context, it is widely believed that sustainable agribusiness ventures can promote job-led economic growth. Agri-preneurship in sustainable agriculture like natural farming, organic farming, eco-friendly agriculture etc. has the potential to transform the face of agriculture in India, the prime driver of our economy. So, there is a demanding need of making suitable policies and frameworks with action plans for achieving the target of agripreneurship. There is a great scope for entrepreneurship in agriculture and this potentiality can be tapped only by effective management of agri-elements such as soil, seed, water and market needs. Agripreneur i.e. an individual with risk bearing capacity and a quest for latest knowledge in agriculture sector can prove wrong the notion of failure and lacking agriculture.

The current study finds a strong need of agripreneurship development in the speciality agriculture. The study compile information on basis agri-entrepreneurship, development of agripreneurship in India along with its scope and need. The study also shows the barriers and challenges for the agripreneur with success factors of the agripreneur that could overpower the challenge. Furthermore, the case studies discussed in the paper makes it clear that speciality agriculture have large scope and area for the agripreneurs to flourish in this particular domain. Also, the paper shows some recommendations or suggestions for the agripreneurs and government of India to excel in the particular field. Hence, this paper has tried to put light on the emerging trends in the field of agripreneurship, its growth and development with respect to the speciality cultivation which needs updation time to time as future endeveour.

References

Alex, J. P. (2011). Powering the women in agriculture: Lessons on women led farm mechanisation in South India. The Journal of Agricultural Education and Extension, 19(5):487-503.

Bairwa, S.L., Lakra, K., Kushwaha, S., Meena, L. K., and Kumar, P. (2014). Agripreneurship development as a tool to upliftment of agriculture. International Journal of Scientific and Research Publications, 4(3):1-4.

Brar, D.S., Kular, J.S., Jasbir, S., and Amrik, S. (2011). Survey of major insect pests of wheat and rapeseed-mustard vis-à-vis control measures adopted by farmers. Journal of Insect Science (Ludhiana), 24(3):235-240.

Chand, K.K. (2019). Agripreneurship: a tool for economic development of India in the new millennium. International Journal on Recent Trends in Business and Tourism (IJRTBT), 3(4):19-25.

Chand, K., Hewa, G., Hassanli, A. and Myers, B. (2020). Evaluation of deficit irrigation and water quality on production and water productivity of tomato in greenhouse. Agriculture, 10(7): 297.

De Wolf, P., and Schoorlemmer, H. (2007). Exploring the significance of entrepreneurship in agriculture. Resarch Institute of Organic Agriculture FiBL, CH-Frick. DOI:10.1142/S1084946706000398

Dollinger, J., Rudi, G., Feurer, D., Belaud, G. and Bailly, J. S. (2003). The use of photogrammetry to construct time series of vegetation permeability to water and seed transport in agricultural waterways. Remote Sensing, 10(12): 2050.

Dubey, S., Pandey, R. K. and Gautam, S. S. (2013). Literature review on fuzzy expert system in agriculture. International Journal of Soft Computing and Engineering, 2(6):289-291.

Gerard McElwee 2006, Farmers as entrepreneurs: Developing competitive skills

Hatfield, J.L. and Walthall, C.L. (2014). Climate change: Cropping system changes and adaptations. Encyclopedia of Agriculture and Food Systems. Elsevier, 256-65.Journal of Developmental Entrepreneurship 11(03):187-206

Kahan, D. (2013). Market-oriented farming: An overview. Food and Agriculture Organization of the United Nations.

Migliorini, P., Galioto, F., Chiorri, M. and Vazzana, C. (2018). An integrated sustainability score based on agro-ecological and socioeconomic indicators. A case study of stockless organic farming in Italy. Agroecology and Sustainable Food Systems, 42(8): 859-884.

Saridakis, G.; Georgellis, Y.; Muñoz Torres, R.I.; Mohammed, A.M. and Blackburn, R. (2021). From subsistence farming to agribusiness and nonfarm entrepreneurship: Does it improve economic conditions and well-being? J. Bus. Res. 136: 567–579. [CrossRef]

Sudharani, V., Vasantha, R., and Supriya, K. (2002). A Study on Profile Characteristics of Rural Young Agri Entrepreneurs. Int. J. Curr. Microbiol. App. Sci, 6(11):252-258.

Uplaonkar, S.S. and Biradar, S.S. (2015). Development of agriculture in India through agripreneurs. International Journal of Applied Research, 1(9):1063-1066.

21

Determinants of Consumer Buying Behaviour for Organic Food Products

Abid Sultan[1], Nikhil[1], Kusumakar Gautam[1], F.A Shaheen[1] Arshad Bhat[2], Irtiqa Malik[1] and Rehana Rashid[1]

[1]School of Agricultural Economics and Horti-Business Management SKUAST-K, Shalimar, Srinagar, Jammu and Kashmir, India
[2]Rajiv Gandhi Chair in Contemporary Studies on Livelihood and Food Security, SKUAST-K, Shalimar, Srinagar, Jammu and Kashmir, India

-

Abstract

The organic food product markets are increasing across the globe, especially given the recent Covid-19 pandemic crisis as well as growing consciousness among consumers about the environment and healthy lifestyle. Organic food products are finding their place in the consumption basket of consumers across the world, including in India. India has a rich traditional and cultural history of organic food products. With the rise of the global industrial revolution, operational limitations along with consumerist culture, the demand for organic food reduced. However, during the last two decades, the rising mindfulness for sustainability in the environment, agriculture, and business practices has again pushed demand for organic products, particularly food products. The rising health awareness among consumers has also added to the growth acceleration of such products. Indian organic food market is projected to be around 100 million USD. Indian organic food markets are slowly expanding in length & breadth, thereby, making India a leading market for organic food products; particularly in metropolitan cities such as Hyderabad, Mumbai, Bangalore, etc. Fundamentally, most of the thoughtfulness in organic food-related products has been on the production side. The consumer psychology side of organic food products has received a low-cut focus. The current chapter describes consumer psychology related to organic food product buying. The chapter identifies significant major antecedents of organic food product buying behavior among consumers.

The current chapter provides buying psychological perspective to the practitioners & entrepreneurs in broadening their understanding, thereby, developing specific interventions & strategies for increasing the spread of purchasing conduct of the consumers for greater sustainability as well as for increasing returns to the farmers for the adoption of organic farming.

Keywords: Organic Food products, Consumer perception, Purchase intentions, UT of JK, Organic Farming, etc.

Introduction

Organic farming is an inclusive system of producing high-quality products through such processes which have the least impairment on the environment, humans, plants, or animal health. Thus, the term in its essence concentrates on sustainability in the farm management practices and food production system. Chopra et al. (2013) in their study defined organic farming as a management & ecological production system that not only promotes biodiversity but also benefits biological cycles and soil nutrition activities. It focuses on minimal use of pesticides and encourages those farm practices that restore balance, and increase ecological harmony. Oroian et al. (2017) through their research advocated that organic farming is an important alternative to conventional farming and solves environmental sustainability issues, and has a long-term positive impact on the soil, water, air, climate changes, biodiversity, limiting the use of genetically modified organisms and encourages the development of ecological services. Literature provides multiple perspectives for organic foods/farming, of which nearly all definitions are based on product attributes such as safety, nutrition, vital qualities, organic integrity, etc. Moreover, these diverse perspectives have one convergence point i.e., usage of the slightest chemical processing and usage of ingredients that are natural & pure in their representation. Literature also defines the word "organic" with other terms such as "bio", "eco", "natural" etc. In recent times, the organic system has received major attention from consumers, policymakers, and practitioners, especially because of the recent Covid-19 pandemic crisis and particularly for organic food products. Organic foods are gaining popularity in both developed and developing countries due to three aspects i.e., nutrition, health, and environment. Many countries across the globe are pushing for the production & marketing of organic food products and are even incentivizing organic farming in their economic system. Countries like the United States of America (USA) have even introduced legal laws (National Organic Program Standards, 2002) for characterizing organic food products. Such initiatives have provided a well-defined framework and standards for defining & describing organic food. The U.S Department of Agriculture (USDA) has also provided guidelines for the description of product labels on organic food products. These labels include

A. **100 percent organic:** This type of label is present on the certified organically grown fruits, vegetables, eggs, meat, and on other single ingredient food products. This type of description is also being used on multi-ingredient foods provided all the ingredients are produced organically and are certified as organic.

B. **Organic:** In the case of multi-ingredient products the label is given to those products having at least 95 percent of certified organic ingredients. The rest non-organic items must be from the list approved by USDA.

C. **Made with organic:** This type of label is provided on products having at least 70 percent of certified organic ingredients. These types of products do not bear a USDA seal.

D. **Organic ingredients:** These products have less than 70 percent of certified organic ingredients. Separate ingredients list provides information about the ingredient whether it is organic or not. These types of products do not also bear a USDA seal.

In the Indian context, the Ministry of Industries and Commerce has implemented National Programme for Organic Production (NPOP) for having a focused and well-directed development plan of action for organic agriculture and quality products. The program proposes the promotion of organic farming and provides an institutional mechanism for the implementation of National Standards for Organic Production and provides information on systems, criteria, and procedures for accreditation of Inspection and certification bodies, the national organic logo. Globally, European Union (EU) is one of the prominent markets for organic food products and occupies the second position in the global organic food product markets, after the United States of America. Their organic farms are bigger than conventional farms. In 2016 total area under organic farming in the EU was 11.9 million hectares which have increased by 18.70 percent as compared to the year 2012 (Wieme *et al.*, 2020). India is an agricultural-driven nation with a huge resource base that can be capitalized on to make India a Hub of organic food products. According to Agriculture & Processed Food Products Export Development Authority (APEDA), India has around 4339184.93 ha (2020-21) area under the organic certification process (registered under National Programme for Organic Production). Out of which 2657889.33 ha is under cultivable area and another 1681295.61 ha for wild harvest collection. Madhya Pradesh is a leading state of India with the largest area under organic certification (i.e. 38.37 percent) followed by Rajasthan, Maharashtra, Chhattisgarh, Himachal Pradesh, Jammu & Kashmir, and Karnataka. Sikkim holds the distinction of having its entire cultivable land (more than 75000 ha) under organic certification.During 2020-21, India produced around 3496800.34 MT of certified organic products which includes all varieties of food products namely oil seeds, fiber, sugar cane, cereals

& millets, cotton, pulses, aromatic & medicinal plants, tea, coffee, fruits, spices, dry fruits, vegetables, processed foods, etc. During 2020-21, India exported around 888179.68 MT of organic produce, which generated foreign revenue of INR 707849.52 Lakhs (1040.95 million USD). The main export destinations of India include the USA, European Union, Canada, Great Britain, Korea Republic, Israel, Switzerland, Ecuador, Vietnam, Australia, etc. Among the different commodities, In terms soya meal (57%) has the highest foreign exchange lead followed by Oilseeds (9%), Cereals and millets (7%), Plantation crop products such as Tea and Coffee (6%), Spices and condiments (5%), Medicinal plants (5%), Dry fruits (3%), Sugar (3%), and others. Though, organic food markets are in the initial stages in India, with most sales coming mainly through exports. The total market size of the organic food market in India is around 100 million USD. The growth of organic food product sales is becoming prominent in the metropolitan cities of India such as Hyderabad, Mumbai, and Bangalore. The total sales of organic products in these cities were about Rs. 9.9 Cr, Rs. 17.9 Cr. and Rs. 21.4 Cr respectively in the year 2007 (Osswald&Menon, 2013). Both central & state governments are making constant efforts to improve institutional support for organic food growers by introducing several policies and programs that are incentivizing organic farming. Several factors are favouring the growing demand for organic food products such as increasing purchasing power, increasing health consciousness, fitness-oriented lifestyle, quality food consciousness, etc. India's organic food market holds a prosperous future in the exports and with the increasing income of the population and growing health concerns.

For fostering the expansion of organic food products in India, especially at the regional levels, it is very imperative to understand the dynamics of consumer purchase intentions along with its determinants for an organic food product. Purchase intentions of organic products is a total of cognitive, affective, and behavioral response of an individual. A consumer purchase intention is a composite process and an outcome of interlinked factors. Organic food products being distinctive products involve a differentiated purchase process from the conventional product buying process. Therefore, it is highly considerable to identify factors of prominence that are influencing the purchase intentions of the consumer viz-a-viz organic food products.

Buying Antecedents for Organic Food Products

The organic food drive started around the 1940s onwards; however, it gained momentum post-1960s. The rising consciousness of consumers for greater sustainability and a healthy lifestyle provided a new pace to its growth and spread. Fundamentally, the scientific technology developments in agriculture proved instrumental in the rise of organic farming among farmers. Also, the rising education level & farmers' conscientiousness provided greater acceptance of the scientific developments & expanded the commercialization of organic foods. Generally, organic food products are characterized as niche products, therefore, providing premium price advantages to the sellers. Also, it provides distinctiveness to the consumers for their purchase and usage. However, the greater side of organic food product purchases entrenches the consumer's psychological perspective. A greater understanding of this was provided by the Theory of Planned Behaviour (TPB) advocated by Ajzen in the 1980s. The theory provided a base for understanding the interaction of the consumer belief system with their behavior. This postulation of TPB aided in fostering ways that can shape consumer belief and perception systems for effective behavior and ultimately strong purchase intentions. According to Spears and Singh (2004), purchase intention is "an individual's conscious plan to make an effort to purchase a product". Using this premise, many studies identified the strong factors that can create strong purchase intentions for organic food products among consumers.

Schlegelmilch *et al.* (1996) intheir research concluded that the market for organic food products is increasing at a fast pace and environmental consciousness plays a significant role and impact on consumer buying behavior towards such products. Similar, observations were also reported by Fotopoulos & Krystallis (2002). They observed that environmental awareness plays a positive role in the organic food purchase of a consumer. Lea & Worsley (2005) findings revealed that the majority of the consumers of organic food products believe that these are healthier, tastier, and better for the environment than conventional food products. They also found that there is a relationship between consumer beliefs and social demographic factors & personal values. Expenses and lack of availability are found to be the two barriers to purchasing organic foods. Results also showed that women perceive organic foods more positively than men and personal values related to nature, environment, and equity positively affects the belief of consumers toward organic food product. Padel& Foster (2005)in their research observed that the motive for the purchase of organic food products includes health consciousness, lifestyle, environmental prosperity, and animal welfare. They also identified some of the purchase barriers such as high price, lack of information/knowledge, lack of confidence, visual product quality & presentation, lack of availability, mistrust of the organic food in supermarkets, eating habits, and convenience needs. Similarly, Chen (2007) in his study found that health and environmental concerns are the two main motives for organic food purchase. The conclusion was also drawn that the attitude of consumers towards organic food products is mainly influenced by their health concerns and their positive attitude towards the environment. Honkanen*et al.* (2006) studied the role of ethical values and motives that drive organic food purchase behavior among consumers. Environment and animal rights perspectives strongly influenced the attitude of consumers towards organic food products. More people are concerned about these issues; a more positive attitude is being developed among the people towards organic food products. Shafie&Rennie

(2009) in their study revealed that the consumer's preference for organic food is highly influenced by an array of the factors like food safety, human health, and environmental concern along with some sensory attributes like nutritive value, taste, freshness, and appearance of the product. The organic food production process is often correlated with the natural process of farming by consumers. Care for the environment, human welfare and the non-use of pesticides and fertilizers attract consumers to these organic food products. While understanding the grounds for increasing organic food consumption, the motivation of consumers towards these products is the most critical factor, which genuinely increases preference & purchase of these products in the market.

Kuhar&Juvancic (2010) observed in their research that purchases of organic food products are greatly influenced by their availability in retail markets, followed by health, and environmental concerns, and the attractiveness of products. Based on this study, they developed a probability model of consumer choice to quantify various determinants of purchase frequency for organically produced fruit and vegetables. Gupta (2009) studied consumer behavior for organic food products in India. He concluded that the quality of food products is one of the important parameters in purchasing decision of the consumer for organic food products. He also concluded that healthiness, cleanliness, freshness, and freedom from pesticides are some of the other primary characteristic motives for consumers to buy organic food products. Chakrabarti (2010) also studied organic food purchases in the Indian context and concluded that health motivation plays a positive influence on the purchase intentions of consumers. The other factors contributing to purchase intentions were the utility of the food, the relevant attitude of consumers in the purchase process, product-specific consumer innovativeness and consumer opinion, and word of mouth.

Ahmad &Juhdi (2010) studied the consumer intentions to purchase organic food products and the results indicate that the intention to buy these products is heavily influenced by the perception and belief of consumers that these products are safe and good for health. Two categories of respondents were observed: buyers and non-buyers. The buyers of organic food products buy these products because they believe that these products are healthier and are better for the environment as compared to conventional food products. Basha et al. (2015) concluded that awareness about any organic food product is the determinant factor, people are aware of the chemicals present in the food items and hence are interested in purchasing organic food products more and more. They also concluded that some prominent factors motivate people to purchase organic food products. These factors include environmental concerns, health concerns, the lifestyle of the consumer, the quality of the product, and subjective norms. These factors influence the purchasing intentions of the consumers also and are one of the motives to purchase these products. Batra (2016) in his study concluded that

Indian consumers have enough consciousness for health and the environment, but still organic food consumption especially in North India is in a fancy stage and can be true to call it immature. Less advertisement and promotion are the main reasons that these products are still in less demand in Indian food markets. Despite being healthy, safe, and eco-friendly, there are misconceptions and distrust among consumers regarding their availability, variety, price, and information about the product and these become the main barriers to purchase and consumption of these products.

Wee *et al.* (2014) advocated that organic food purchase intentions are significantly influenced by consumer perception of safety, health, environmental concerns, and animal welfare of the product. The actual buying behavior of the people about the product is significantly affected by the purchase intentions of the product. Chandrashekar (2014) observed in his research that consumer behavior affects the organic food product markets in the world and marketers need to be very innovative and dynamic to compete with the changing purchase behavior for organic food products, especially among urban residents. Marketers must create realistic & innovative promotions to create a successful market demand for such products. Besides, quality, health, nutritional value, taste, freshness, and general appearance such as the color of the product also affect consumers' purchase intentions. Dash et al. (2014) stated that the increased use of chemicals in farming has shifted consumer tastes & preferences towards organic food products, which has led to the rise of domestic as well as global demand for such products. Awareness and knowledge about these products are crucial in changing the attitude and behavior of consumers towards these products, which ultimately increases the growth of organic food markets. Socioeconomic variables also play an important role in building a positive attitude of consumers towards organic food products. It was seen that self-decision and friends play a vital role in the purchase decision-making process of organic products.

Singh (2017) in his research concluded that knowledge and awareness about organic food products are two important aspects that create health concerns among people and there is a shift in consumption patterns of organic food products among different age groups. Significant changes occur due to advertisement and promotion which lead to increased awareness and consumption of healthy foods. The organic food markets increase all over the world which leads to increased investment opportunities and expansion and promotion of small-scale farmers and small and medium scale industry. Due to environmental sustainability concerns, the consumption pattern of people shifted towards organic food products from conventional food products, which are available in limited amounts though there is a specific demand for these products. Among different organic food products vegetables, fruits and beans are highly demanded and the price of vegetables especially leafy vegetables is more as compared to the price of conventional vegetables. He also advocated that the increased health consciousness among people plays an important role in gaining attention towards these products and the preferences of the consumer are mainly affected by the quality of the product which includes health perspective, nutritional value, taste, freshness, and general appearance of the product. Consumers' intention to buy the product is also influenced by the supply of the product and there is an imbalance in supply and demand and less access to information about these products and which ultimately affects the willingness of consumers to purchase the product.

Sahu& Mishra (2018) investigate consumer preferences about organic food. Rapid urbanization and growing consumer's concern for food safety, health, and the environment in recent times have resulted in a new tide in the consumption of organic foods. It was observed that the awareness and consumption of organic foods are more in developed countries than in developing countries. The health benefits of organic food products are well known and this has increased the demand for these products. In countries like India, where organic food markets are still in the nascent phase of their growth, the level of awareness and acceptance is relatively low and hence this affects the purchase intentions of organic food products. Bagher et al. (2018) observed that promoting consumer behavior to purchase organic food products as the element of a green marketing perspective can be a fundamental way of increasing the social responsibility among the people and consequently reducing environmental damage. It was shown that organic knowledge, health awareness, and environmental concerns had a positive effect on both perception and intentions to buy the product. A healthy lifestyle, subjective norms, perceived behavior control, and ethical orientation of people towards such products also showed positivity in intentions to purchase organic food products.

The review of the literature reveals that consumer buying decision-making is an important process influencing consumer purchase behavior. The literature reveals an array of factors that contributes to developing consumer perception about the purchase of a product. In the context of organic food products the prominent factors contributing to the perception of the consumer includes environmental concern, food safety concern, health consciousness, food taste, advertisement and promotions, brand image, the lifestyle of the consumer, quality of the product, and ethical consideration. These factors greatly influence the purchase intentions of a consumer for organic food products. Based on the review of prominent literature about organic food products, consumer perception, consumer behavior, and purchase intentions, the present study identified five prominent parameters related to consumer perception related to organic food products i.e., Environment & ethical consideration, health & food safety concerns, quality & taste, brand image, and advertisement & promotion. A brief description of each of the identified variables is as:

1. **Environmental & Ethical Consideration:** It contributes towards primary purchasing motivation cues for the consumers. The preference for products grown organically reduces environmental hazards, reduces pollution, and uses energy-efficient production systems. Besides, it also favors those business/farming practices and processes that are environmentally friendly and ethical in their operational existence.
2. **Health & Food Safety Concerns:** Contemporary consumers are highly information driven. The information used in the purchasing decision process has grown over the years, which has resulted in a change in the purchase behavior of consumers. With more awareness and knowledge about the production & processing of food products, health consciousness, and food safety concerns have increased among the consumers, resulting in more preference and adoption of such food which has high health benefits and more assured food safety consumption.
3. **Advertisement and Promotion:** The increasing adoption of organic food products requires greater advertising and promotion strategies for leveraging upon the triple benefits associated with organic products i.e., economic, social, and environmental benefits. These benefits need to be disseminated among the people for having organic consumerist culture.
4. **Quality and food taste:** Organic food products possess a distinctive character & different tastes and more importantly their consumption is different from traditional food products. Therefore, quality & taste assurance must be guaranteed in organic food markets. The role of quality certifications and standards should be done within an inclusive framework.
5. **Brand image:** Brand is an intangible asset that creates distinctiveness within consumer minds. Organic food products being different from traditional products need to be crafted, developed, and strengthened through a strong brand image, reflecting the distinctiveness, benefits as well as niche recognition to the consumers.

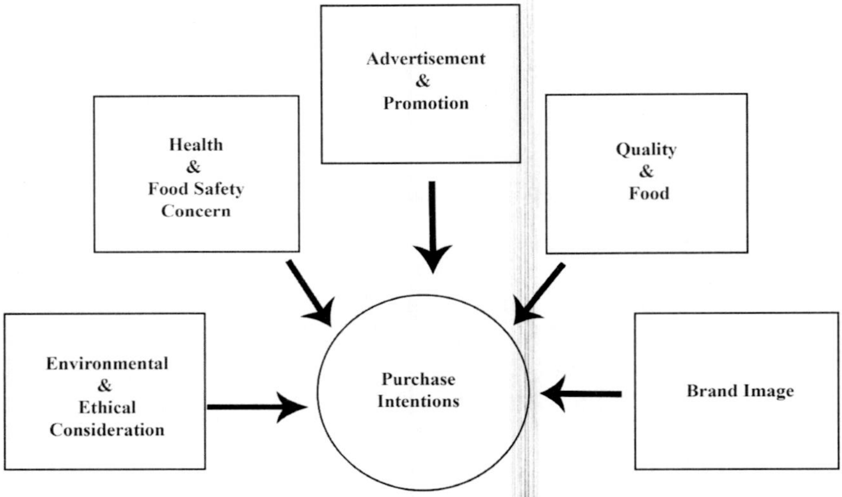

Fig. 1. Conceptual Framework

The identified five parameters can be instrumental in strengthening the purchase intentions of the consumers. The current chapter provides a theoretical perspective on the development of organic food marketing.

Conclusion

The current study provides for comprehending the consumer perspective attached to the purchasing of organic food products. The study identified five major determinants that have the potential of influencing consumer purchase intentions for organic food products. The study can be beneficial for understanding the organic food purchase buying behavior among consumers. It can also provide thoughtful cues to the policymakers, practitioners, and entrepreneurs in understanding elements of purchase behavior for organic food products as well as the significance of building positive consumer perception for having favorable purchase intention behavior.

Acknowledgment: The present chapter is a part of the master-level thesis conducted at the School of Agricultural Economics and Horti-Business Management, Shere-e-Kashmir University of Agricultural Science and Technology of Kashmir, J&K, India

References

Ahmad, S. N., & Juhdi, N. (2010). Organic Food: A Study on Demographic Characteristics and Factors Influencing Purchase Intentions among Consumers in Klang Valley, Malaysia. International Journal of Business and Management, 5(2), 105-118.

Chopra, A., Rao, N., Gupta, N., and Vashisth, S. (2013). Come Sunshine or Rain: Organic Foods Always on Tract: A Futurist Perspective. International Journal of Nutrition, Pharmacology, Neurological Diseases, 3(3), 202-205

Basha, M. B., Mason, C., Shamsudin, M. F., Hussain, H. I., & Salem, M. A. (2015). Consumer Attitude Towards Organic Food. Precedia Economics and Finance,31 (15), 444-452.

Batra, R. (2016). Consumer Behavior and Purchase Intention in Indian Organic Food Markets. Amity Business Review, 17 (2), 60-76.

Bagher, A. N., Salati, F., & Ghaffari, M. (2018). Factors Affecting Intention To Purchase Organic Food Products Among Iranian Consumers. Academy Of Marketing Studies Journal, 22 (3), 1-23.

Chen, M.F.(2007). Consumer Attitudes and Purchases Intentions in Relation to Organic Foods in Taiwan: Moderating Effects of Food Related Personality Traits, Food Quality & Preference, 18(7), 1008-1021

Chakrabarti, S. (2010). Factors Influencing Organic Food Purchases in India-Expert Survey Insights, British Food Journal, 112(8), 902-905

Dash, M., Samantaray, A., & Dash, M. (2014). Consumer's Perception Towards Organic Food. International Journal Of Engineering And Management, 4 (3), 110-113.

Fotopoulos, C., & Krystallis, A. (2002). Purchasing Motives and Profile of the Greek Organic Consumer: A Countrywide Survey. British Food Journal, 104 (9), 730-765.

Gupta, K. B. (2009). Consumer Behavior for Food Products in India. Annual World Symposium of the Indian Institute of Management, Bombay, India, 20-21.

Honkanen, P., Verplanken, B., & Olsen, S. O. (2006). Ethical Values and Motives Driving Organic Food Choice. Journal of Consumer Behavior: An International Research Review, 5(5), 420-430.

Kuhar, A. and Juvancic, L. (2010), What Determines Purchasing Behavior For Organic and Integrated Fruits and Vegetables, Bulgarian Journal of Agricultural Sciences, 16(2), 111-122

Lea, E., & Worsley, T. (2005). Australian's Organic Food Beliefs, Demographics, and Values. British Food Journal, 107 (11), 885-869

Oroian, C. F., Safirescu, C. O., Huran, R. Chiciudean, G. O., Arion, F. H. &Muresan, L. C. (2017). Consumer's Attitude Towards Organic Products and Sustainable Development: A Case Study of Romania. Sustainability, 9 (9), 1559-1573

Osswald, N., & Menon, M. K. (2013). Organic Food Marketing in Urban Centres of India. International Competence Centre for Organic Agriculture (ICCOA), 1-47. Accessed on 14-08-2022, available at https://d-nb.info/1220733911/34

Padel, S., & Foster, C. (2005). Exploring The Gap Between Attitudes And Behaviour Understanding Why Consumers Buy Or Do Not Buy Organic Food. British Food Journal, 107 (8), 606-625.

Singh, J. (2017). To Study the Awareness of Consumers Towards Organic Products. International Journal of ManagementResearch and Business Strategy, 6 (1), 188-194.

Sahu, R.K. and Mishra, R.K (2018). Consumer Preferences on Organic Food Products: A Review. Asian Journal of Multidimensional Research (AJMR), 7(3), 19-28

Schlegelmilch, B. B., Bohlen, G. M., & Diamantopoulos, A. (1996). The Link Between Green Purchasing Decisions and Measures of Environmental Consciousness. European Journal of Marketing, 30 (5), 35-55.

Shafie, F. A., & Rennie, D. (2009). Consumer Perceptions Towards Organic Food. Procedia-Social and Behavioral Sciences, 49 (2012), 360-367.

Spears, N., & Singh, S. N. (2004).Measuring attitude toward the brand and purchase intentions. Journal of Current Issues & Research in Advertising, 26(2), 53-66

Wee, C. S., Ariff, M. S., Tajudin, M. N., & Zakuan, N. (2014). Consumers Perception, Purchase Intention and Actual Purchase Behavior of Organic Food Products. Integrative Business and Economics, 3 (2), 378-397.

Wieme, R. A., Reganold, J. P., Crowder, D. W., Murphy, K. M., & Carpenter-Boggs, L. A. (2020). Productivity and Soil Quality of Organic Forage, Quinoa, and Grain Cropping Systems in the Dryland Pacific Northwest, USA. Agriculture, Ecosystems and Environment, 293 (2020), 1-13.

22

The Status of Cold-water Fisheries and Prospects of Their Utilization for Poverty Reduction in Jammu and Kashmir

Sahar Masud

Fisheries Unit, Sher-e-Kashmir University of Agriculture Sciences and Technology-Jammu, Jammu and Kashmir, India

The Himalayas, which cover 594 400 km2, run for about 2 500 km from west to east, between Nanga Parbat (8126 m) in the west and Namcha Barwa (7756 m) in the east with three parallel ranges namely The Great or Inner Himalayas, The Lesser Himalaya or Himachal and TheShiwaliks (Fig.1).The fish species distribution in the Himalayan streams depends on the flow rate, nature and substratum, water temperature, and the availability of food.This region has a dense net of rivers and streams, as well as numerous lakes, manmade reservoirs, floodplains and swamps, man-made ponds and irrigation canals of cold-water character. These water resources have a diverse fish fauna that support the livelihoods of rural communities within the region. They also provide various opportunities for development of cold-water fisheries and aquaculture.

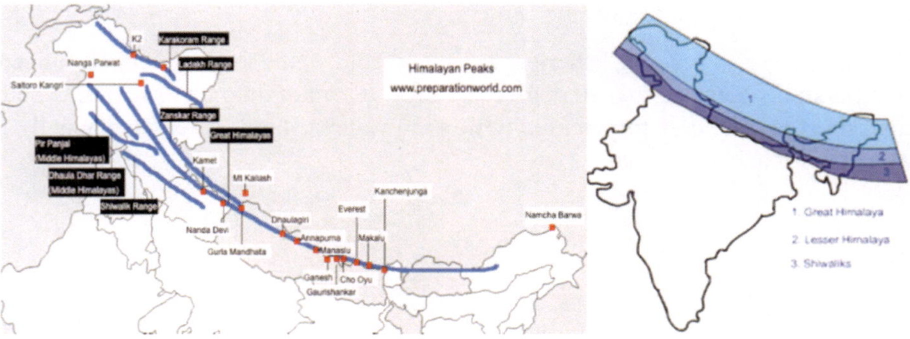

Fig. 1. The Himalayan Mountains

The Jammu and Kashmir state, situated in the extreme northern part of the country with a total geographical area of 2,416 thousand hectares is divided into four agroclimatic zonesviz, Temperate Zone, Sub-Temperate (Mid Zone), Sub–Tropical Zone and Cold Arid Zoneeachhaveitsownspecificgeo-climaticcondition, whic hdeterminesthecroppingpatternandproductivity. Over the years, agriculturists and farmers have adopted several area-specific and time-specific cultivation practices to meet therequirementoftheirstaplefoodcrops.There is currently a shift towards cultivating low volume high-value cash crops.Fisheries also play an important role in providing food and income to the mountain people.

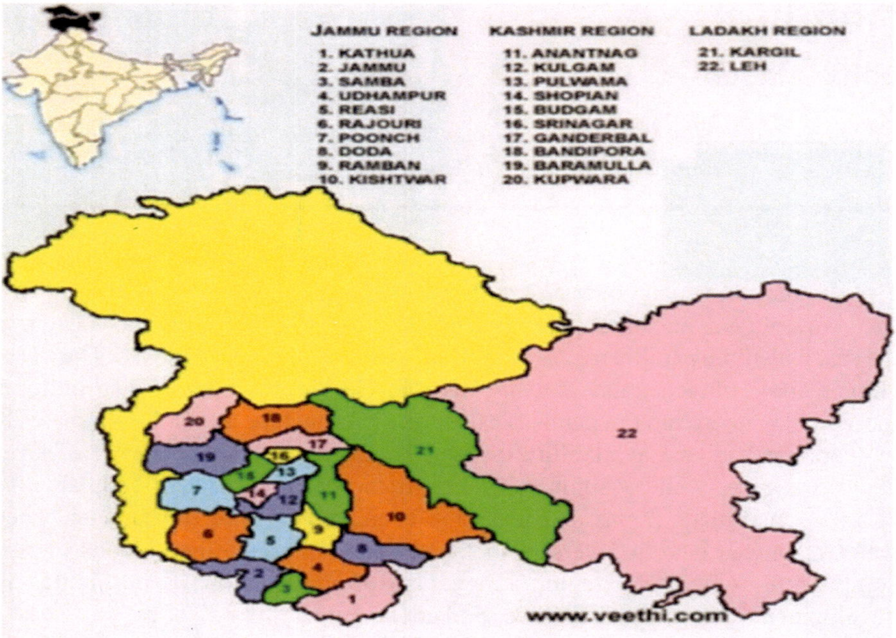

Fig. 2. Map of Jammu and Kashmir

Kashmir has two types of fisheries, warm and cold water. For survive, trout needs temperatures between 0 to 20 degrees Celsius. It cannot flourish in standing water and must have access to perennially flowing waters. It also migrates annually to the upper reaches to breed.

Fig. 2. Agro-climatic Zone of Jammu and Kashmir

Table 1. Major characteristics of different agro-climatic zones of J&K

S.N.	Particulars	Jammu		Kashmir	Ladakh
		JK-1 (sub-tropical Zone)	JK-2 (Intermediate/ Sub-Temperate (Mid Zone))	JK-3 (Temperate Zone)	JK-4 Temperate Zone)
1	Geographic aldistribution	Jammu District, parts of Kathua district, parts of Rajouri district, parts of Udhampur district and lower areas of Reasi) , parts of Doda District (Ramban)	Parts of Kathua District, Parts of Poonch District (Surankote), parts of Rajouri District (Rajouri and Kalakote), parts of Udhampur District parts of Reasi, some portion of Doda District, Ramban and Kishtwar), besides minor locations in Uri Tehsil of Baramulla District and Karnah Tehsil of Kupwara district.,	Allsix districts of Kas hmirvalleyviz., Anantnag, Pul wama, Srikangar, Budgam, Parts of Poonch, Rajouri, Doda, Kathua and Udhampur Districts of Jammu Division	Twodistricts of Ladakh (Leh, Kargil)
4	Averagel and holdings	0.99	0.93	0.53	1.08
5	Netirrigated area (%)	36	10	62	100
6	Majorrivers	Ravi,Tawi	Chenab	Jhelum	Indus,Shyok
7	Altitude (mslrange)	300-1350	800-1500	2400-3000	3500-8400
8	Averag eannual rainfall	1069	1649	789	83
9	Temperature (^{0}C)				
	Minimum	13.6	11.5	1.2	-7.0
	Maximum	>32.1	31.4	24.5	17.4
10	Thermal	Mild	Mild	Cold	Verycold
11	Hydricindex	Humid	Humid	Humid	Arid

Among the available natural resources, excellent quality of water is abundant in the J&K mountainous regions (Fig.3), where income-generating and employment opportunities are scarce, cold-water fish farming could help to ensure employment and steady incomes. Therefore, the utilization of available water resources for cold-water fish production is an obvious possibility. In addition to the production, cold-water fisheries could also ensure increased income and employment through angling tourism, restaurants, tourism and related services.

Fig. 3. Fishery Resource Map of Jammu and Kashmir

The man most associated with the introduction of trout in Jammu & Kashmir (J&K) was the redoubtable angler, and owner of a carpet factory, Frank J Mitchel. Although an indigenous variety – the snow trout – already existed in the waters, Mitchel convinced Maharaja Pratap Singh, to ask the British for more varieties. In 1899 the Duke of Bedford sent a consignment of 10,000 trout eggs to J&K, but they perished on the way. A second shipment arrived from Scotland in 1900. Some of these were released in Dachigam, while the rest were reared by Mitchel until they were fingerlings, and then released in other streams of the Valley. The rainbow and brown trout adapted well to the J&K, while the snow trout continued to flourish. Since then, trout have flourished in Kashmir. Hence, In J&K there are three variety of Trout.

Brown trout(*Salmo trutta fario*)attracts anglers, particularly foreigners, to J&K's high-altitude lakes and numerous snow-fed freshwater streams, making it an important part of the tourism industry. Unfortunately, a blend of pollution, human intervention and climate change has resulted in a decline in the brown trout's population. So, it is now also reared in state-run farms and introduced into streams and rivers to keep them populated.

Rainbow trout(*Oncorhynchus mykiss*) is predominantly blue, green, yellow and silver in colour, and glistens in various shades of the rainbow. It is rich in protein and is said to boost immunity, another reason for its popularity. But it faces threats when raised in open water sources. The breeding of the rainbow trout is disturbed due to mineral extraction and other activities. That is why it is now reared in ponds, with a defined entry and exit of running water.The rainbow trout is undisputedly the most widely introduced and cultured salmonid fish across the globe. It is an extremely popular food and sport fish that thrives in highly oxygenated cold freshwater, which is naturally abundant in the mountainous regions of India. However, over the years, the exotic rainbow trout has become the most relevant and remunerative cold-water fish farmed in Indian uplands garnering more significance as a source of employment and food security. The biological characteristics that make rainbow trout truly appealing for aquaculture are rapid and high growth potential, ease of captive breeding, robustness of fry at first feeding, ability to withstand high density culture conditions without compromising with growth and welfare standards.

Snow Troutor *Schizothorax sp.* once a Kashmir's endemic fishis now vanishing due to pollution and introduction ofthe exotic speciesi.e., Common Carp and Trout, which were introduced in 1956 and 1900 have been thriving at Schizothorax's cost.It was found in abundance in snow-fed lakes, rivers, wetlands and canals until 30 years back. They have elongated and subcylindrical fleshy bodies and are known to be tastier and free of diseases. Their presence in water bodies would indicate that they were pollution free.Experts say over half of Schizothorax's 12 species have disappeared. Research in 2005 found 15% of the total catch to be Schizothorax while 68% was Common Carp. In 2015, Schizothorax's numbers reduced to 10%.

Mahseers (*Tor sp.)* Mahseer is acclaimed as a world famous, outstanding game and food fish of India. As a sport fish, it provides unparalleled recreation to anglers from all over the world, better than salmon. It is known as tiger in waters, because of the fight it musters to wriggle off the hook. In the past mahseer formed a substantial natural fishery in the major riverine and lacustrine have been a legendary sport fish of India with a very high table value. Mahseer, at one time considered to be of single species, is now represented by six valid species distributed all over India. Despite their abundance, at one time mahseer are reported to be declining in size and numbers and are feared to be in danger of extinction in some parts of the country ecosystems of India. In commercial fisheries it occupies an important position for its good quality. For the fishermen mahseer is of considerable importance because of its large size. As a food fish, it is highly esteemed and fetches the highest market price in north and northeast of India.

Major threats to the Cold-water Fisheries

- The real attraction of the cold-water fisheries is in the wild, in the numerous snow-fed fresh water streams. Unfortunately, it is in the wilderness that the cold-water fisheries is declining due to a deadly blend of pollution, human intervention and climate change.
- There are multiple causes for these problems.The use of biocides (pesticides, insecticides, herbicides, fungicides, etc) and other chemicals in the horticulture and agriculture activities have contaminated our water resources. Consequently, the aquatic biota, particularly fishes are getting affected.
- Hydroelectric power projects (HEPP), barrages, weirs are also immensely responsible for the decline in cold-water population in Jammu and Kashmir waters.cold-water is a migratory fish and during their upstream and downstream migration, they are often obstructed andalso exposed to hazards due to the passage of turbines from hydropower plants. In the case of downstream migrations, this potential hazard, characterized by the turbine mortality rate, depends to a significant extent on the turbine model, the operating parameters and the type and size of the fish.
- Scientists project that for trout and other cold-water aquatic species, the combined effects of a 1.5° Fahrenheit increase in average air temperatures could result in a 7 to 16 percent loss of trout habitat in the continental United States (Steven 2008).

Strategies for Scaling up cold water fishculture and Production

The total trout production in Jammu and Kashmir is 600 tonnes annually generating a revenue of more than Rs 13 crore. Kashmir regi on is the major contributor in trout production. The state hasatotal of 0.3 lakh hectares of in land water bodies in the form of 27781 km of rivers and canals, 0.7 lakh hectares of reservoirs, 0.17 lakh hectares of tanks and ponds, and 0.06 lakh hectares of flood plain lakes/ derelict waters (Ayyappan *et.al.*, 2011).The rivers and streams are suitable for trout farming. In addition to riversand streams, troutcan be cultured in race ways taking water from rivers, stream sand even lakes. Realizing the potential of trout farming, there is a need of technological, administrative and financial improvement to increase the production of trout.

Multi-pronged strategies are needed to stimulate horizontal and vertical enhancement of trout production in India. Especially, the productivity of existing operations should be augmented by ushering in scientific advances made in cold-water farming and marketing. The focus should be on:

- Expansion of trout farming in feasible areas with correct choice of new farm sites and best use of water in the most economical way.
- Establishment of cold-water fish hatcheries in private sector
- Development and recommendation of appropriate extension strategies for production enhancement.
- Linkages among research institutes at national, regional and international level for the advancement of research support to trout growers/industry.
- Commercialization of cold-water fisheries through public-private partnership (farmers' cooperative, NGOs etc.) approach:
    - Expand domestic and explore abroad market for the cold-water fisheries.
    - Develop input delivery mechanism in trout growing pocket areas.
    - Initiatives on establishment of fish feed industry in private sector – Encourage to local manufacturers
    - Preparation database on fish production, demand, consumption.
    - Development of training curriculum for farmer level training
    - Development of human resources, infrastructure, facilities required for the rapid expansion of trout
    - Develop/strengthen feedback mechanism and communication system for all stakeholders
- Development of code of conduct for responsible aquaculture with approach of:
    - Water user's right
    - Farm Registration
    - Water Resource conservation
    - Effluent water monitoring
- Research needs
    - Genetics and breeding – Genetic improvement, Brood management (Fecundity improvement, Maturity alteration, Alevins survival)
    - Year-round supply of trout seed.
    - Feed and nutrition – Feed standardization, Digestibility improvement, Cost effective feed
    - Health management – Regular monitoring for site specific protocol, Research ties up with other research fields.

- Production research – Economic harvest size, Stocking density versus water flow, Alternative production systems
- Post-harvest – Value addition, transportation (Quality maintenance)
- Socio –economic - Consumer preference, Marketing cost.
- Environmental – Effluent discharge, Effluent utilization
- GIS based evaluation on potential sites of cold-water fish / Site characterization – GIS and Remote sensing
- Development of technology on Low-cost Pond/raceway construction

- **Improvement of Research facilities**
  - Upgrading of lab facilities
  - Seed research facilities
  - Production research facilities

23

Mustard Economics of KVK Beneficiaries and Non-beneficiaries in Samba District of Jammu Division

Tsewang Dolma, Vinod Gupta, Sudhakar Dwivedi Ruhana Rafiq and Chanchal

Division of Agriculture Extension Education, Faculty of Agriculture Sher-e-Kashmir University of Agricultural Sciences and Technology of Jammu, Jammu and Kashmir

Abstract

KVK (Krishi Vigyan Kendra) is a unique extension system to provide solutions to farmers and operate under the control of state agricultural universities or central institutes situated in a particular area. There are about 20 KVKs established in Jammu and Kashmir, out of which nine KVKs are under the Sher-e-Kashmir University of Agricultural Sciences and Technology Jammu, Chatha. The main objective of KVKs is to demonstrate the latest technologies through trainings and create awareness among the farmers as well as workers through front Line Demonstrations. In this line the present study was conducted during 2019-20 to study mustard economics of KVK beneficiaries and non-beneficiaries. Primary data was collected through structured questionnaire and analysis of data was carried through Z test and t test. The sample of the study was based on random sampling technique to obtain a total sample size of 120 mustard growers (60 beneficiary farmers and 60 non-beneficiary farmers). Result shows that average seed cost of the beneficiaries and non-beneficiaries were 411.66 and 310.83 Rs./ha, the average cost of fungicide of the beneficiaries and non-beneficiaries were 291.41 and 171.46 Rs./ha for Albugo occidentalis and Erisiphe cruciferam, average cost of urea of the beneficiaries and non-beneficiaries were 614.33 and 574 Rs./ha. The study also shows that total cost of the beneficiaries and non-beneficiaries were 311173.67 and 29596.33 Rs./ha. The B.C ratio of the beneficiaries and non-beneficiaries were 1.97:1 and 1.34:1, respectively.

Keywords: KVK, Cluster, Frontline Demonstrations, Extension activities, Training, Adoption, Impact.

Introduction

Agriculture is the foundation of India's economy, as it provides food to people, feed to cattle, raw materials to primary and secondary industries, contributes upto 19 percent of GDP and employs 54.6 percent of the workforce (Kirti and Prasad, 2016). Due to diverse soil and climate conditions, vast varieties of crops such as cereals, pulses, spices , millets, oilseeds, flowers and fruits are being grown in India. Among these crops grown, India is among the leading producers of pulses, cereals and spices in the world (Mishra *et al.*, 2021). India is the fourth largest oilseed economy in the world. Among oilseeds, there are nine crops that are the primary sources of oils in the country, out of which seven are edible oils (soybean, rapeseed-mustard , groundnut, sunflower, sesame, niger and safflower) and two are non-edible (linseed and castor). The above mentioned oilseed crops are grown throughout the year, and mustard is the sole oilseed crop which is well-suited to rabi season. It is a key source of revenue and contributes significantly to the livelihood security of resource to poor, small and marginal farmers living in rain-fed areas of the country (Shekhawat *et al.*,2012). Mustard plant belongs to the Brassicaceae family and Brassica or Sinapis genera (Swati *et al.*, 2015). Mustard seeds are high in oil and protein content and its oil content ranges from 24 to 40 per cent and protein content ranges from 17 to 26 per cent. The seeds of mustard are mostly processed for the extraction of oil, and the residue left behind known as mustard cake, is used for cattle feed (Kumrawat and Yadav, 2018). Mustard aphid (*Lipaphis erysimi*(L.) is the most notorious, cosmopolitian louse-like and obligate ectoparasite which causes a bulk of the qualitative and quantitative loss of rapeseed-mustard crops (Koirala, 2020). Timely and efficient crop management practices such as, disease pest resistant varieties along with adoption of proper crop rotation, selection of high yielding, timely planting, adequate plant stand, balanced plant nutrition, need base plant protection, irrigation and timely weed control have great influence on productivity of mustard. Package and practices for mustard are developed by agricultural universities, research institutes and are refined, demonstrated, popularised through Krishi Vigyan Kendras (KVK) and field extension departments in different states etc. Krishi Vigyan Kendra in particular from 1992 onwards have been on the frontline in providing farmers with wide variety of services like frontline demonstrations, technology testing/ refinement, vocational training and popularising concepts of integrated farming system, entrepreneurship, crop diversification, value addition, biodiversity conservation, organic farming etc (Katole *et al.*, 2017). There are about 20 KVKs in Jammu and Kashmir, out of which nine KVKs are under the administrative control of Sher-e-Kashmir University of Agricultural Sciences and Technology, Jammu (SKUAST –J Chatha). Empirical evidences have highlighted that KVK system has highly and positively impacted the rural farming community in terms of yield, income, productivity, sale price and above all capacity for optimal utilization of resources etc (Singhal and Vatta, 2017). In this line, this study was carried during the year

2019-20, to assess the impact of KVK's cluster frontline demonstrations on mustard overall economics, cost of cultivation and income of farmers in Samba district of Jammu division. Cluster frontline demonstrations mainly involves selection of farmers through interactive discussions/meetings, organizing these farmers into groups (10-15 per group), identifying the gaps in adoption of currently existing package of practices, conducting frontline demonstration of newly developed or refined package of practices on farmers field, supply of required inputs in the form of high yielding seed varieties, insecticides, fertilizers, weedicides etc, regular monitoring and technical support by KVK experts and at last impact assessment in terms of yield, productivity, selling price, quality etc. (Jha *et al.*, 2020).

Methodology

Ex-post facto design was applied to conduct the study and multi stage sampling plan was followed for the selection of ultimate farmers. Samba district was selected purposively because KVK, Samba has laid maximum number of Front Line Demonstrations (FLDs) under its Cluster Frontline Demonstration Programme. A list of 166 mustard growing beneficiaries was collected from KVK Samba. Out of the available list, 60 number of mustard growing beneficiary were selected by using random sampling technique. Equal number of non-beneficiary mustard growing farmers were also selected from the representative/adjoining villages. By this way total sample size comprised of 60 beneficiary and 60 non-beneficiary mustard growing farmers. Data was collected at individual farmer level, through open ended questionnaire particularly on economics, cost of cultivation and of the mustard crop and income of the farmers along with data on adoption practices, socio economics and constraints faced in mustard crop production. Data was tabulated and analyzed through SPSS software.

Result and Discussion

Economics of mustard crop

This result depicts the various costs incurred in the usage of inputs, cultivation of mustard crop like cost of seeds, cost of fertilizers, cost of ploughing, human labour costs etc. Further, it also reveals the net income, gross income and on farm and off farm income of the farmers of the study area.

Cost of Cultivation of Mustard Crop

Table 1 shows the various cost incurred in the usage of inputs and cultivation of mustard crop. It revealed that 411.66 and 310.83 Rs./ha of seed cost was incurred in the category of beneficiaries and non-beneficiaries, respectively with difference value of 100.83. This might be due to the reason that beneficiaries were growing more mustard crop on large fields as compare to non-beneficiaries. Further, expenditure incurred in usage of insecticide for the management of *Lipaphis erysimi* and *Athalia lugens proxima* was found highest in the category of beneficiaries (402.83 Rs./ha) as compare to non-beneficiaries (293.33 Rs./ha) with difference value of 109.5.. Among insecticide used for the management of *Lipaphis erysimi*, insecticide dimethoate was used, which have expenditure value of 641 and 510 Rs./ha in the category of beneficiaries and non-beneficiaries, respectively with difference value of 131, whereas, for the management of *Athalia lugens proxima,* cypermethrin was used, which have highest expenditure value in the category of beneficiaries (164.66 Rs./ha) as compare to non-beneficiaries (75.83 Rs./ha) with difference value of 88.83. As far as fungicide usage is concern for the management of *Albugo occidentalis* and *Erisiphe cruciferam*, expenditure incurred in usage of fungicide was found highest in the category of beneficiaries (291.41 Rs./ha) as compared to non-beneficiaries (171.46 Rs./ha) with difference value of 119.95. Among fungicides used for the management of *Albugo occidentalis*, ridomil-mz was used and have expenditure value of 193.33 Rs./ha in the category of beneficiaries and 270 Rs./ha in the category of non-beneficiaries with difference value of 76.67, whereas, carbendazim was only used by beneficiariesand have expenditure value of 36 Rs./ha. For the management of *Erisiphe cruciferam,* ridomil-mz was used and have highest expenditure value in the category of beneficiaries (296.66 Rs,/ ha) as compare to non-beneficiaries (293.33 Rs./ha) with difference value of 3.33, whereas, carbendazim was also used and have highest expenditure value in the category of beneficiaries (639.66 Rs./ha) as compared to non-beneficiaries (141 Rs./ha) with difference value of 498.66. Further, expenditure value of herbicide pendimethalin was found highest in the category of beneficiaries (62 Rs./ha) as compared to non-beneficiaries (30 Rs./ha) with difference value of 32. This might be due to the reason that the area of mustard crop for the beneficiaries wrere more for beneficiaries. With regard to fertilizer usage, average cost incurred in the usage of urea was highest in the category of beneficiaries (614.33 Rs./ha) as compare to non-beneficiaries (574 Rs./ha) with difference value of 40.33, whereas, 1198.66 and 1390.66 Rs./ha was found as average cost of DAP in the category of beneficiaries and non-beneficiaries, respectively with difference value of 192. Similarly, 315.5 and 172.66 Rs./ha was found as average cost of MOP in the category of beneficiaries and non-beneficiaries with difference value of 142.84. With regard to land preparation, ploughing cost was incurred highest in the category of beneficiaries (10783.33 Rs./ha) as compared to non-beneficiaries (10283.33 Rs./ha) with difference value of 500. Further, cost of harvesting was incurred highest in the category of non-beneficiaries (9200 Rs./ha) as compared to beneficiaries (8900.00Rs./ha) with difference value of 300. Cost of threshing was incurred highest in the category of non-beneficiaries (3241.66 Rs./ha) as compare to (3128.50 Rs./ha) with difference value of 113.16.The other human labour cost was incurred highest in category of beneficiaries (2800.00 Rs./ha) as compared to non-beneficiaries (2500.00 Rs./ha) with difference value of 300.00. The total cost incurred by comparising various cost involved in mustard was found highest in the category of beneficiaries (31173.67 Rs./ha) as compared to non-beneficiaries (29596.33 Rs./ha) with difference value of 1577.34. The difference in total cost of cultivation of beneficiaries and non- beneficiaries might be due to the reason that the respondents of beneficiaries were using resources more intensively and were spending higher amount on all the inputs namely urea, DAP, MOP application of herbicides and pesticides and other human labour cost.Yield of mustard crop was found highest in the category of beneficiaries (12.25 q/ha) as compared to non-beneficiaries (8.06 q/ha) and the result of mustard yield was found significant at 1 per cent level of significance with difference value of 4.19.The reason behind this might be that the beneficiaries had cultivated KVK, Samba recommended varieties (high yielding varieties) which yields twice the yield obtained by the traditional mustard variety which is desi-sarson. The rate of selling of mustard crop was found highest in the category of beneficiaries (5026 Rs./q) as compared to non-beneficiaries (4920 Rs./q) with difference value of 106. Beneficiaries and non-beneficiaries have gross income of 61576.67 and 39688.00 Rs. with difference value of 21888.67, whereas, the net income of beneficiaries and non-beneficiaries was found as 33203.00 and 12547.00 Rs., respectively with difference value of 20656.00. So, the B:C ratio of beneficiaries and non-beneficiaries was found as 1.97:1 and 1.34:1, respectively. which means that the supremacy of recommended package of practices under the frontline demonstration for beneficiaries was more profitable over the farmers practices of non-beneficiaries. Above findings were in line with Chaudhary *et al*. (2018).

Table 1: Operation Wise Cost of Cultivation of Mustard Crop (Rs./ha)

Particulars	Beneficiary (n=60)	Non-Beneficiary (n=60)	Difference
Seed cost(Rs./ha)	411.66	310.83	100.83
Pesticide use in mustard crop			
Insecticide cost (Rs./ha)			
Dimethoate(*Lipaphis erysimi*)	641.00	510.00	131.00
Cypermethrin(*Athalia lugens proxima*)	164.66	75.83	88.83
Overall	402.83	293.33	109.50
Fungicide cost (Rs./ha)			
Ridomil-Mz(*Albugo occidentalis*)	193.33	270	76.67
Carbendazim(*Albugo occidentalis*)	36.00	0.00	36.00
Ridomil-Mz(*Erisiphe cruciferam*)	296.66	293.33	3.33
Carbendazim(*Erisiphe cruciferam*)	639.66	141.00	498.66
Overall	291.41	171.46	119.95
Herbicide cost (Rs./ha)			
Pendimethalin	62.00	30.00	32.00
Fertilizer use in mustard crop			
Average cost of urea (Rs./ha)	614.33	574.00	40.33
Average cost of DAP (Rs./ha)	1198.66	1390.66	192.00
Average cost of MOP (Rs./ha)	315.50	172.66	142.84

Particulars	Beneficiary (n=60)	Non-Beneficiary (n=60)	Difference
Land preperation cost			
Ploughing cost (Rs./ha)	10783.33	10283.33	500.00
Cost of harvesting (Rs./ha)	8900.00	9200.00	300.00
Cost of threshing (Rs./ha)	3128.50	3241.66	113.16
Other human labour cost	2800.00	2500.00	300.00
Total cost(Rs./ha)	31173.67	29596.33	1577.34
Yield(qtl/ha)	12.25	8.06	4.19
Rate per quintal (in Rs.)	5026.00	4920.00	106
Gross income (in Rs.)	61576.67	39688.00	21888.67
Net income (in Rs.)	33203.00	12547.00	20656.00
B.C Ratio	1.97:1	1.34:1	

Off-Farm and On-Farm Income of Respondents

Table 2 depicts that total income was found highest in the category of beneficiaries (409756.03 Rs.) as compared to non-beneficiaries (283110.70 Rs.) with difference value of 126645.33. Among these, beneficiaries have off farm income of 356569.70 Rs. and non-beneficiaries have off farm income of 249613.70 Rs. with difference value between these two categories was 106956.00. Similarly, beneficiaries have on farm income of 53186.33 Rs. and non-beneficiaries have on farm income of 33497.00 Rs. with difference value between these two categories was 19689.33.

Table 2: Distribution of Income from Off- Farm and on Farm

Income (Rs.)	Beneficiary	Non-Beneficiary	Difference
Off farm income	356569.70	249613.70	106956.00
On farm income	53186.33	33497.00	19689.33
Total income	409756.03	283110.70	126645.33

Conclusion

Beneficiary farmers were getting higher productivity as well as good revenue from mustard crop as compared to non-beneficiary farmers as the beneficiary farmers were practicing all the package of practices that were demonstrated to them by the KVK experts during their field visits and cluster frontline demonstrations. In order to increase the benefit of CFLDs in horizontal way, lead farmers need to be further motivated to transfer the skill and knowledge to the fellow farmers in their vicinity through group discussion, panchayat sabhas etc. Further research station technology needs to be further popularised through extension programme as it has the potential of improving production of mustard at state level without increasing in area. Also the extension agencies in collaboration with KVKs should held demonstrations at regular interval so as to show the effects of new technology in mustard production and motivate farmers for adoption of new technology to bridge the yield gap between lab and land. Moreover, Krishi Vigyan Kendra in the state need to play more active and lead role in providing proper technical support to the farmers through different educational and extension activities to reduce the extension gap for better cereal, pulses, oilseeds, spices and fruit crop production in the state.

References

Chaudhary, R.P., Choudhary, G.K., Prasad, R., Singh, R. and Chaturvedi, A.K. 2018. Impact assessment of front line demonstration on mustard crop. International Journal if Current Microbiology and Applied Sciences, Special issue-7: 4737-4742.

Jha, A. K., Chatterjee, K., Mehta, B. K. and Kumari, M. 2020. Assessment of technological interventions on productivity and profitability of chickpea (cicer arietinum l.) through Cluster Frontline Demonstrations (CFLDs) in Sahibganj district of Jharkhand. International journal of advanced biological research, 10(4): 213-217.

Katole, S. B., Bhatt, J. H and Patel, G. G (2017). Impact analysis of activities of Krishi Vigyan Kendra. Gujarat Journal of Extension Education, 28(2): 267-270.

Kirti, R and Prasad, S. 2016. FDI impact on employment generation and GDP growth in India. Asian journal of Economics and Empirical Research, 3(1): 40 -48.

Koirala, S. 2020. Mustard aphid and crop production. International Journal of Applied Sciences and Biotechnology, 8(3): 310-317.

Kumrawat, M and Yadav, M. 2018. Trends in Area, Production, and Yield of Mustard crop in Bharatpur Region of Rajasthan. IJEDR, 6(1): 315-321.

Mishra, P., Yonar, Y., Yonar, H., Kumari,K., Abotaleb,M., Das, S.S. and Patil, S.G. 2021. State of the art in total pulse production in major states of India using ARIMA techniques. Current Research in Food Science, 4:800-806 https://doi.org/10.1016/j.crfs.2021.10.009.

Shekhawat, K., Rathore, S.S., Premi, O. P., Kandpal, B. K. and Chauhan, J. S. 2012. Advances in Agronomic Management of Indian Mustard (Brassica juncea (L.) Czernj. Cosson): An Overview. International journal of Agronomy, 2012, 1-14. https://doi.org/10.1155/2012/408284.

Singhal, S and Vatta, V. 2017. Impact of Krishi Vigyan Kendra on adoption of improved agricultural production practices. International Journal of Science, Environment and Technology,6(2): 993-1000.

Swati., Sehwag, S and Das, M. 2015. A brief overview: Present status on utilization of mustard oil and cake. Indian Journal of traditional medicine, 14(2): 244-250.

24

Forecasting of Trends in Production and Import of Edible Oil in India using ARIMA Model

P. Kanakarani[a] and J. Gayathri[b]

[a]*Sri Sarada College for Women (A), Tamil Nadu, India*
[b]*Bharathidasan University, Tamil Nadu, India*

Abstract

India is the second largest consumer of edible oil in the world next to China and world's leading importer of edible oil. The import of edible oil consumes considerable foreign earnings. After, 1990s changes in economic policy and shift in consumer behaviour in demanding food items including edible oils. It was observed that import of edible oil was steadily increasing towards upwards from 1995. This study aims to compares the relation between trend in area cultivated, production and productivity of edible oilseeds with import of edible oil in India from 1980-81 to 2019-20. Secondly, this study forecasted the production of edible oilseeds and import of edible oil for next ten years using ARIMA model. Auto Correlation Function (ACF) and Partial Auto Correlation Function (PACF) are calculated for the time series dataset. This was fitted with Box Jenkins ARIMA model for area under cultivated (0,1,2), production (0,1,1) and productivity (0,1,1) of oilseeds in India and import (1,1,1) of edible oil in India. The forecasted results revealed that nearly 27.08 million hectares of land is required for cultivation of oilseeds with upper and lower limits of 30.52 million hectares from 23.67 million hectares. The forecasted production was 36.86 million tonnes with the lower limit 30.69 to the upper limit of about 43.03 million tonnes in the year 2029-30. The forecasted import was about 20.7 million tonnes with upper and lower limits of 264 million tonnes from 151 million tonnes respectively.

Keywords: Oilseeds, Area Under cultivation, Production, Productivity, Import, Forecasting, ARIMA, CAGR.

Introduction

There are forty different edible oilseeds which can be used for edible and non-edible purposes. Out of which nine oilseeds are extensively used for edible and non edible purposes. The oilseeds such as groundnut (peanut), sunflower, sesame, soybean, rapeseed-mustard, niger and safflower are generally used for edible purposes and castor and linseed are used for non-edible purposes. Plants are the main source for 71 percent of edible oil. Some oil crops such as groundnut (peanut) and sesame can be used directly as food and edible oil, but others are mainly processed to obtain oil, and oil cake or meal (Hetje, 1989). The oil crops are grown all over the world (Girish Kumar, 2012). Area under cultivation of edible oilseeds in 2020 was 27.14 million hectares. The total production was 33.22 million tonnes with the productivity i.e. yield per hectare of 1025 kilogram per hectare in India (Ministry of Statistics). India is the world's largest importer of edible oil (Santosha, et. al., 2017) and the import of edible oil in 2020 was 93.37 million tonnes.

Review of Literature

Ajoy Kumar et.al (2017) study, towards oilseeds sufficiency in India: present status and way forward, reveals that due increasing percapita consumption of edible oil would increase the demand for oil seeds in India. The gap between domestic production and demand for oil seed is extending in India. The gap between domestic production of oilseeds and demand is widening as the result of failure to adopt higher yield of oilseeds for cultivation (Ramasamy and Selvaraj 2002:302). They suggested that increasing the domestic production of oil seeds is necessary to control import of edible oil in India. John et.al (2006) study of Groundnut trade liberalization: Could the South help the south?, they concluded that policies are affecting global groundnut product markets. Their analysis revealed that African exporters would gain through reductions in subsidies in India and to a least through China. Narasinga Rao (1992) study reveals that palm oil as an important importing edible oil in India. Importing palm oil from Malaysia helps to meet the shortage of edible oil in India. At the same time the low-cost palm oil would stimulate higher level of consumption and helps to reduce energy deficiencies among the young children. The 66th NSS Round, Report No. 514 reveals that monthly per capita edible oil consumption in India was 636 gram in rural area and 818 gram in urban area.

Erik et al (2003:3) asserted that India is the second largest consumer in the world, closely behind China and world's leading importer of vegetable oil since 1998. Two-third of domestic demand in India is met by imported edible oil (Rajesh Bhayani, 2020:2). In 2019 India imported edible oil worth of Rs. 7,300 crores for 15 million tonnes. This alone account for 40 percent from total agricultural commodities import bill and three percent from overall import bill in India. From total imported edible oil 62 percent was palm oil imported from Indonesia and Malayia followed by this 21 percent of soya oil imported from Argentina and Brazil and 16 percent sunflower oil imported from Ukrain and Argentina.

Number of studies used Compound Annual Growth Rate (CAGR) to capture growth and instability of production, productivity and yield. Sajeev Kumar (2020) used CAGR to measure growth performance of oilseeds production in Rajasthan. Chandra Mohan (2017) measured area, production and productivity of groundnut in India using CAGR. Ankur (2018) used to analyse rice production. Birhanu and Sekar (2016) used compound annual growth rate to find out the trends and instability in maize production.

Forecasting technique is used to find out the performance of past and based on this to predict future trend. ARIMA is widely used tool for forecasting with the help of time series data (Santosha et. al., 2017). If the time series data set is linear ARIMA is most widely used (Ravichandran et al., 2018). Studies related domestic and export of agricultural production is mostly using ARIMA model for forecasting (Amin et al., 2014, Celik, 2017) applied ARIMA model for time series data in order to forecast annual amounts of honey production from the years 1950 and 2015 in Turkey.

This paper aims first, to compare the trend in area cultivated, production and productivity of edible oil seeds with import of edible oil in India from 1991 to 2020 by using CAGR. Secondly, to forecast area, production and productivity of edible oilseeds and imports of edible oil in India for the next ten years the ARIMA model is applied.

Limitations of the Study

This study is based on secondary data. The analysis of edible oil seeds limited the factors of imports and domestic production, productivity and area under cultivation.

Research Design

The research design is based on analytical approach. The literature reviewed and published data from Ministry of Statistics are the main sources of data. The time series dataset for the periods from 1991 to 2020 for area production (in million tonnes), productivity (in kg/hectare) and imports (in million tonnes) are used to trace out the trend and to forecasting. **Estimation of Trend Line: Annual growth rate** are calculated in order to fix the trend line in respect of area, production, productivity and imports. This helps to find out the relation between domestic production of edible oilseeds with import of edible oil. The index values are estimated by fixing 1991 as base year. ARIMA model was separately constructed for the variables of area under cultivation of oilseeds, production, productivity and import of edible oil in India for the years from 2021 to 2030 using SPSS package. **The ARIMA Model** of output for variables are listed below with model identification. a.) Area ARIMA (0,1,2) Model –I, b.) Production ARIMA (0,1,1)) Model –II, c.) Productivity ARIMA (0,1,1) Model –III and d.) Import ARIMA (1,1,1) Model –IV.

Estimation of Annual Compound Growth (CAGR) : The annual Compound Growth was estimated by using following formula.

$$\text{CAGR} = \left[\frac{\text{CV}}{\text{PV}}\right]^{1/n} - 1$$

CV = Current year value, PV = Previous year Value and n= Number of years.

Results and Discussion

Trend in Annual Growth Rate of Import of Edible Oil and Area, Production, and Productivity of Edible Oil Seeds in India

The introduction of High Yielding Varieties of Seeds Programme helped to record a considerable progress in food grain production in India. In the case of "other" crops like edible oil seeds, pulses and coarse seeds the performance has not been so good. These "other" crops have an important place in the Indian agriculture covering an area of about 26.5 million hectares, with a total production of over 29 million tonnes in the triennium ending 2011-12. This constitutes 14.8 percent of the gross cropped area in the country (Vijay Paul Sharma, 2017:3). During 2011-12, India imported 9.2 million tonnes of edible oils which was about half of its domestic requirement. Oilseed cultivation is undertaken across the country in about 26 million hectares on marginal lands, dependent on monsoon rains nearly 72 percent of area under oilseeds is rain fed and with low levels of input usage (GOI, 2012-13:104).

Table 1: Index of Area under Cultivation, Production and Productivity of Oilseeds and Import of Edible Oil in India from 1991-2020

Year	Area (in Million tonnes)	ACGR (%)	Production (in Million tonnes)	ACGR (%)	Productivity (in Kg/hectare)	ACGR (%)	Import (in Million tonnes)	ACGR (%)
1991	24.15	0	18.61	0	771	0	0.53	0
1992	25.89	0.7	18.60	0	719	-0.7	0.22	-8.42
1993	25.24	-0.25	20.11	0.78	797	1.04	0.10	-7.58
1994	26.90	0.64	21.50	0.67	799	0.03	0.11	0.96
1995	25.30	-0.61	21.34	-0.07	843	0.54	0.35	12.27
1996	25.97	0.26	22.11	0.35	851	0.09	1.07	11.82
1997	26.31	0.13	23.33	0.54	887	0.42	1.42	2.87
1998	26.10	-0.08	21.32	-0.9	817	-0.82	1.27	-1.11
1999	26.20	0.04	24.75	1.5	945	1.47	2.62	7.51
2000	24.26	-0.77	20.71	-1.77	854	-1.01	4.20	4.83
2001	22.77	-0.63	18.44	-1.15	810	-0.53	4.18	-0.05
2002	22.64	-0.06	20.66	1.14	913	1.2	4.32	0.33
2003	21.49	-0.52	14.84	-3.26	691	-2.75	4.37	0.12
2004	23.66	0.97	25.19	5.43	1064	4.41	5.29	1.93
2005	27.52	1.52	24.35	-0.34	885	-1.83	4.50	-1.6
2006	27.86	0.12	27.98	1.4	1004	1.27	4.29	-0.48
2007	26.51	-0.5	24.29	-1.4	916	-0.91	4.27	-0.05
2008	26.69	0.07	29.76	2.05	1115	1.99	4.90	1.39
2009	27.56	0.32	27.72	-0.71	1006	-1.02	6.71	3.19
2010	25.96	-0.6	24.88	-1.07	958	-0.49	8.03	1.81
2011	27.22	0.48	32.48	2.7	1193	2.22	6.67	-1.84
2012	26.31	-0.34	29.80	-0.86	1133	-0.51	8.45	2.39
2013	26.48	0.07	30.94	0.38	1168	0.3	11.01	2.68
2014	28.05	0.58	32.75	0.57	1168	0	10.45	-0.52

Year	Area (in Million tonnes)	ACGR (%)	Production (in Million tonnes)	ACGR (%)	Productivity (in Kg/hectare)	ACGR (%)	Import (in Million tonnes)	ACGR (%)
2015	25.60	-0.91	27.51	-1.73	1075	-0.83	12.73	1.99
2016	26.09	0.19	25.25	-0.85	968	-1.04	15.64	2.08
2017	26.18	0.03	31.28	2.16	1195	2.13	14.01	-1.1
2018	24.51	-0.66	31.46	0.06	1006	-1.71	15.03	0.7
2019	24.79	0.12	31.52	0.02	937	-0.71	15.55	0.34
2020	27.14	0.91	33.22	0.53	1025	0.91		

Source: Computed from secondary data, www.indiastat.com

From table 1, there was an inverse trend movement was observed between import of edible oil and production and area under cultivation of oilseeds. At the same time positive relationship was observed cultivated area oilseeds and its production. The area under cultivation and production oilseeds and import of edible oil were stood at 24.15 million hectares, 18.61 million tonnes and 0.53 lakh tonnes respectively. The area cultivated, production of edible oilseeds and import of edible oil increased to 27.13 million hectares, 33.22 million tonnes and 15.5 million tonnes in 2019-20. The import was jumped at an average of 1.19 million tonnes, 5.09 million tonnes and 12.17 million tonnes in the last three decades.

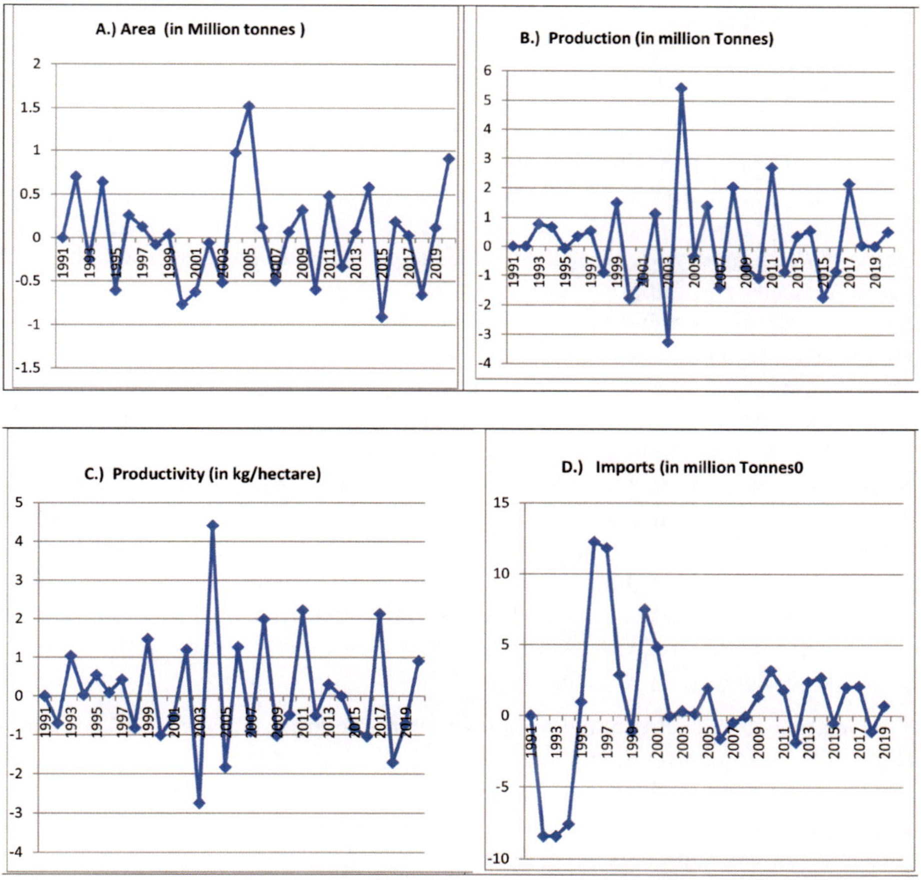

Fig. 1. Trend in Area, Production, Productivity and Import of Edible Oil in India

The trend lines based on ACGR for area under cultivation of oilseeds, production, productivity and import of edible oil in India are depicted in Fig. 1. The trend line of import is dramatically moving towards upwards from 1995 onwards. The import of edible oil trend is closely associated with domestic production of oilseeds. During this period area has been reduced considerably followed by production and there was not much improvement in productivity which responses for increase in import of edible oil in India. At the same time from 2004 to 2006 the import of edible oil was reduced considerably which was closely associated with area and production of edible oilseeds in the same period. From the fig. 1 it is also observed that the percentage in annual growth rate of production is closely resembled with productivity than area under cultivation. Oilseeds productivity in India is comparatively less than the major oilseeds producing countries in the world. According World Scenario of vegetable oil and fats for 2013-14 groundnut productivity in India is nearly three times less than USA and its productivity in India was 1642 kg per hectare and the world average was 1777 kg per hectare.

Forecasting Area, Production, Productivity oilseeds and Import of Edible oil in India

Time series analysis has been applied to observe the performance of area under cultivation, production, productivity i.e. yield of edible oilseeds and import of edible oil in India during the period from 1991 to 2019. It also explains the changes associated with area under cultivation, production and productivity of edible oilseeds and import of edible oil for over the period of time. A series of data forecasted separately for area under cultivation, production, productivity of edible oilseeds and import of edible oil in India for next ten years. Thus, a list of observations of a sequence of time series data representing that the measurements are taken at regular intervals. In this study ARIMA model used to forecast the trend of area under cultivation, production , productivity of edible oilseeds and import of edible oil in India based on the past performance of the respective variables time series dataset. ARIM model is an acronym which of consists of three components namely an autoregressive (AR) model, an integration (I) model and a moving average (MA) model. ARIMA model is widely used to forecast the future values of a dataset. This study aims to find out the growth rate for the next ten years from 2021 to 2030. The steps involved in constructing an ARIMA model are presented below.

To construct an ARIMA model for a given variable time series dataset is required to produce an autocorrelation function (ACF) and a partial autocorrelation function (PACF). The ACF and PACF are required to decide the order of autoregressive and moving average. In case the data are not stationary then the order of differentiation is used to convert stationary. These are elaborated in ARIMA model through p,d.q.

First. the constructed ACF and PACF are shown in figures 2 and 3. If the ACF spikes remain significant or more lags which shows that data are not stationary. In such a case the first order difference (d=1) transformation is applied to convert non-stationary dataset into stationary dataset.

Secondly, the processes of auto-regression are determined by exponentially declining ACF and spikes in the first one or more lags of the PACF. The order of auto-regression is determined by number of spikes. In this ACF and PACF such kind of scenario was not observed therefore, it can be inferred that the auto-regressive processes is zero (p=0).

Thirdly, the processes of moving average are determined by exponentially declining PACF and spikes in the first one or more lags of the ACF. The order of auto-regression is determined by number of spikes. In this ACF and PACF such kind of scenario was not observed therefore, it can be concluded that the moving average processes is zero which is denoted as q=0.

Auto Correlation Function (ACF) and Partial Autocorrelation Function (PACF) for Area under Cultivation, Production and Productivity of Edible Oilseeds and Import of Edible Oil in India

The analysis of ACF and PACF for the variables of cultivated area, production and productivity of edible oilseeds are compared with import of edible oil in India for the next ten tears i.e. 2021-30. The measured values of autoregressive integrated moving average model order p,d, q are helps to determine ARIMA functions of variables. Results of estimation are presented in table 2.

Table 2: Goodness of Fit Measures of ARIMA Model for Oilseeds Production and Import of Edible Oil

Fit Statistics	Mean			
	Area under cultivation	Production	Productivity	Imports (old data based on 1981)
Model	ARIMA (0,1,2)	ARIMA (0,1,1)	ARIMA (0,1,1)	ARIMA (1,1,1)
Stationary R-squared	0.124	0.35	0.446	0.048
R-squared	0.251	0.636	0.480	0.952
RMSE	1.431E3	3.069E3	104.124	11.312
MAPE	4.661	9.658	8.577	78.839
MaxAPE	13.182	62.034	30.379	720.294
MAE	1.181E3	2.201E3	81.706	8.902
MaxAE	2.833E3	9.205E3	209.916	25.386
Normalized BIC	14.881	16.291	9.523	5.209
L-Jung-Box Q(18)				
Statistics	18.004	18.175	18.111	17.093
Significant	0.324	0.378	0.382	0.380

Source: Computed from table 2

The application of this model is to find out the systematic pattern in residuals of the model . If it is there will be removed to improve chosen ARIMA. This is done by examining ACF and PACF of the residual and L-Jung Box statistics (Refer table 2). These are given in the figure 2 and 3. The ARIMA model no-significant static indicates goodness of "fit" model.

Stationary R –Squared: This is a measure which compares the stationary part of the model to a single mean model. The stationary R-squared is positive and highly significant to the ARIMA (1,1,1) of imported edible oil followed by ARIMA (0,1,1) models of production and ARIMA (0,1,1) of productivity of edible oilseeds in India. It was ranges between 0.382 to 0.378. It can be concluded that increasing area and productivity of edible oilseeds would help to control the import in the near future.

R-Squared: The R-squared fit index indicates variability of a time series dataset. The measured R- squared value is interpreted below for each model. Highest variability was observed in imports of edible oil and production of edible oilseeds with the R-squared values of 0.952 and 0.636. The variability of area under cultivation of oilseeds was 0.251 which is low as compared with productivity 0.480. It is observed that controlling the variations in area and production oilseeds would help to control the increasing variations in import of edible oil in the near future.

Root Mean Square Error (RMSE)

This fit index helps to find out the range of difference between the dependent time series dataset and its predicated values of the time series. The obtained RMSE value 104.124 and 11.312 are highest for the ARIMA model (0,1,1) of productivity and ARIMA (1,1,1) model of importing edible oil. The RMSE values for production and area under cultivation was 3.069 and 1.431. So it can be concluded that the ARIMA model for these variables are perfectly fit (Refer table 2)._

Mean Absolute Percentage Error (MAPE) : The fit statistics of MAPE for oilseeds cultivated area, production and productivity is less turns out to be 4.661, 9.658 and 8.277 respectively. These obtained results revealed that the forecasting accuracy is high. On the other side the forecasting accuracy is also good for imports of edible oil which is estimated at 78.839. There was not much change in forecasted area under cultivation. It was stable. On the other side productivity and production of oilseeds shows an increasing trend. The import of edible oil also shows increasing trend. This is due to the import was continuously moving upwards but the fluctuation is closely resembled with production.

Normalised BIC: Normalised Bayesian Information Criterion fit index also known as ARIMA model index explains the overall fitness of the model. The normalised BIC value is depends on the mean square error, length of the time series dataset and other parameters. The forecasted values of normalised BIC indicates that the fitness of the ARIMA model. The observed value of normalised mean is least which indicates that the predicated ARIMA model is good. The obtained normalised BIC was lowest for import of edible oil which is 5.209. It was highest for production of oil seeds followed by area under cultivation and productivity 16.291, 14.881, and 9.523 respectively.

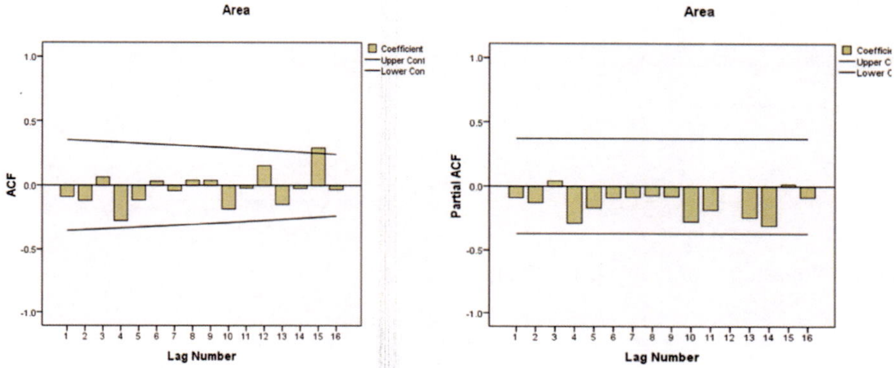

Fig. 2.1. Area Under Cultivation of Edible Oilseeds in India: ACF &PACF

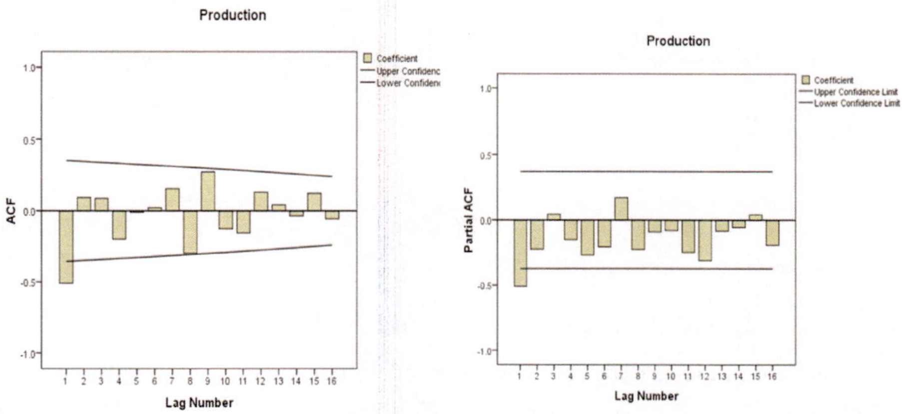

Fig. 2.2. Production of Edible Oilseeds in India: ACF &PACF

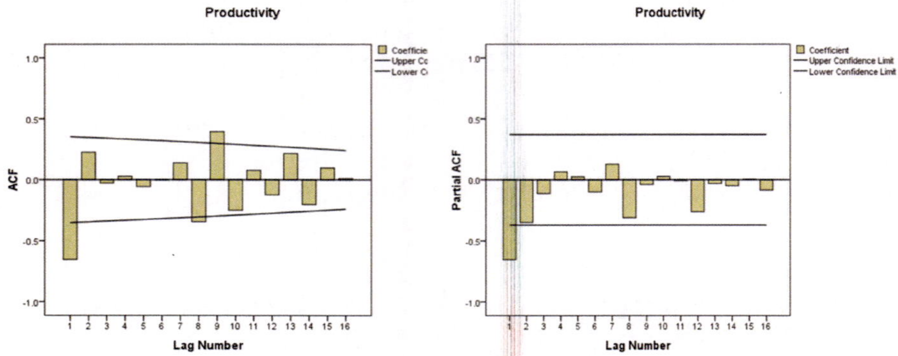

Fig. 2.3. Productivity of Edible Oilseeds in India: ACF &PACF

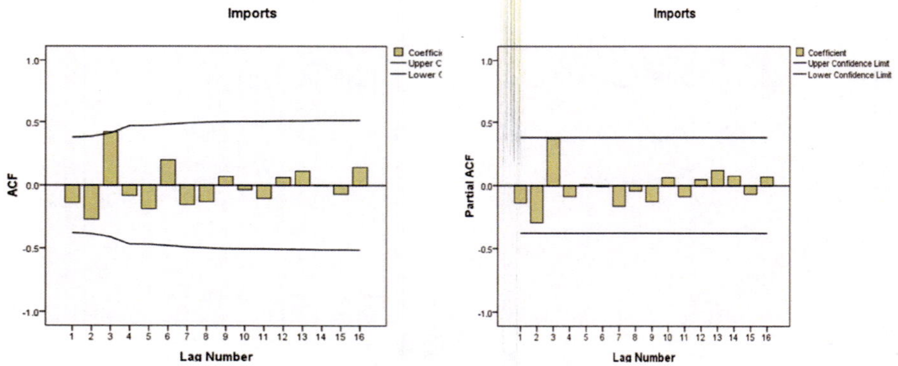

Fig. 2.4. Import of Edible Oil in India: ACF &PACF

L-Jung-Box Q and Residuals of ACF and PACF

The estimated models "good fit" was observed with the help of L-Jung Box Q and residuals of ACF and PACF. The estimated L-Jung Box statistics for the models of the import of edible oil and edible oil seeds cultivated area, production and productivity given in table 2. And the fitness is also associated with residuals of fitted ARIMA models given in figure 3.1, 3.2, 3.3, and 3.4.

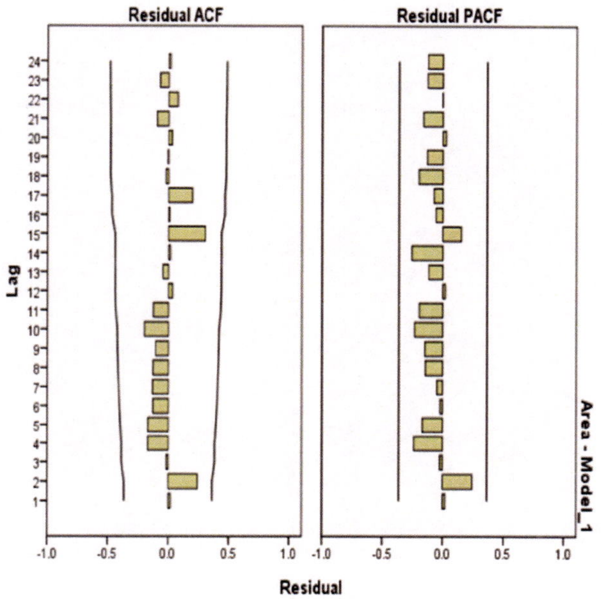

Fig. 3.1 ACF and PACF Residuals of Fitted ARIMA model for Area under Oilseeds inn India

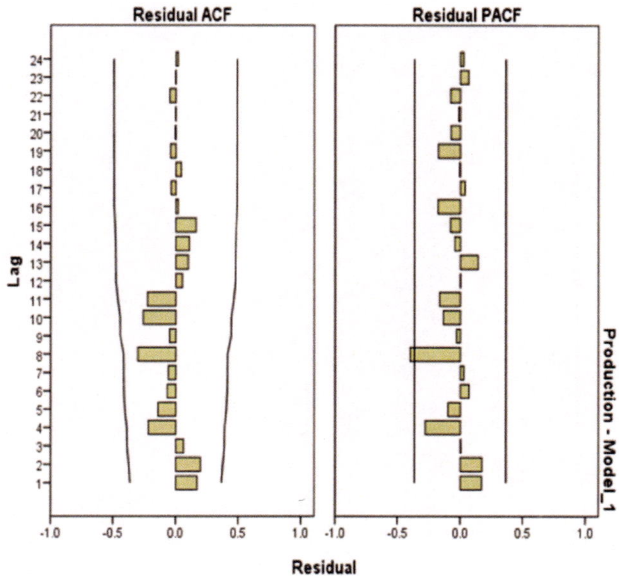

Fig. 3.2 ACF and PACF Residuals of Fitted ARIMA model for Production of Oilseeds in India

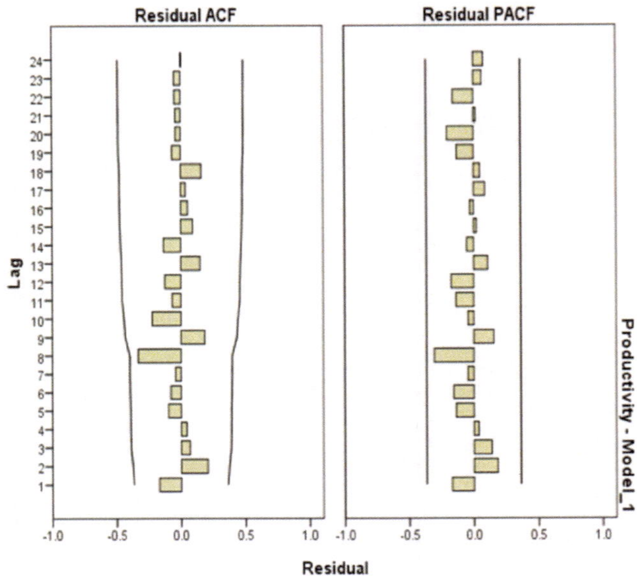

Fig. 3.3 ACF and PACF Residuals of Fitted ARIMA model for Productivity of Oilseeds in India

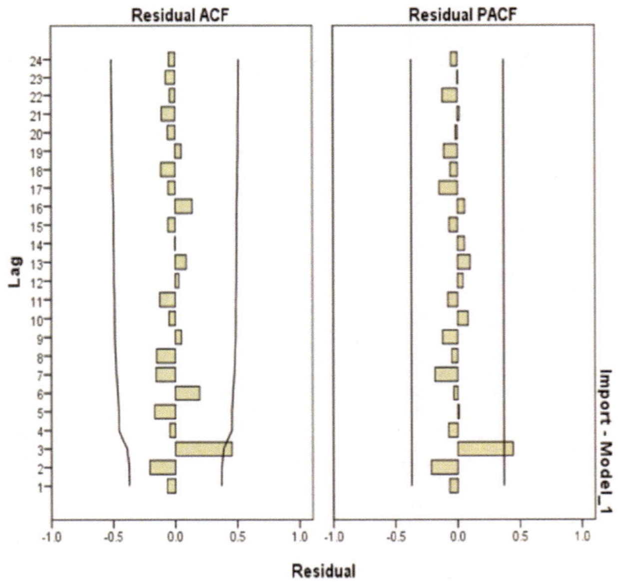

Fig. 3.4 ACF and PACF Residuals of Fitted ARIMA model for Import of Edible Oil in India

Table 3: Forecasted Area Under Cultivation, Production, Productivity of Edible Oil Seeds and Import of Edible Oil

	Forecast					Lower Control Limit (LCM)			Upper Control Limit			
Year	Area	Production	Productivity	Import	Area	Production	Productivity	Import	Area	Production	Productivity	Import
2020-21	27283	32585	1079	156	24448	26413	866	133	30117	38757	1292	178
2021-22	26638	33060	1089	164	23209	26888	870	131	30066	39232	1308	197
2022-23	26695	33535	1099	169	23266	27364	874	132	30123	39707	1325	206
2023-24	26752	34010	1110	175	23323	27839	878	134	30180	40182	1341	215
2024-25	26809	34486	1120	180	23381	28315	883	137	30236	40656	1357	223
202526	26866	34961	1130	186	23438	28790	887	139	30293	41131	1374	232
2026-27	26923	35436	1141	191	23496	29266	892	142	30349	41605	1390	240
2027-28	26980	35911	1151	196	23553	29742	897	145	30406	42080	1405	248
202829	27037	36386	1162	202	23611	30218	902	148	30462	42554	1421	256
2029-30	27094	36861	1172	207	23669	30693	907	151	30518	43028	1437	264

Source: Results of ARIMA Model

Forecasting of Edible Oilseeds area cultivated, production and productivity and the import of edible oilseeds in India

The ARIMA model has best fit indices which are helps to compare the all possible variables and therefore the values predicated form 2020-21 to 2029-30 are presented in table -2.

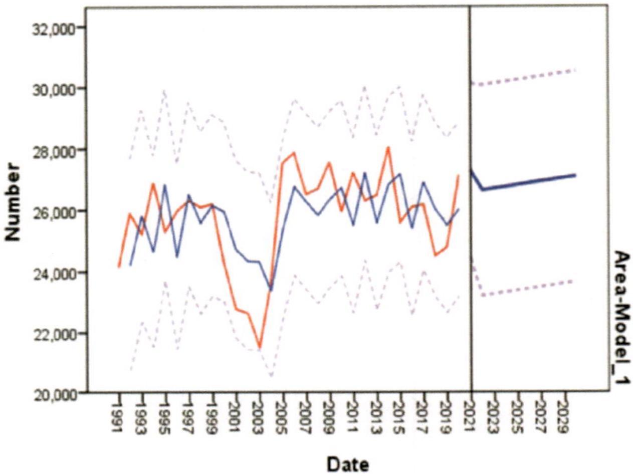

Fig. 4.1. Series Chart with forecasted values of Area Under Cultivation of Edible Oil Seeds in India from 1991 -92 to 2029-30

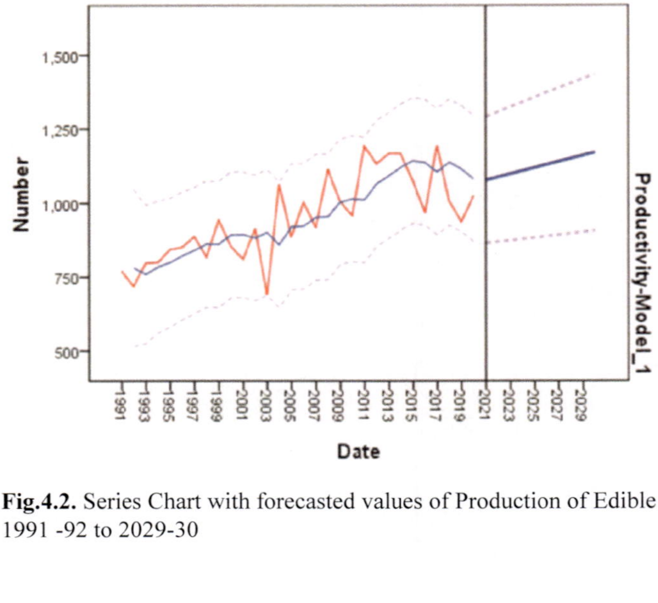

Fig.4.2. Series Chart with forecasted values of Production of Edible Oil Seeds in India from 1991 -92 to 2029-30

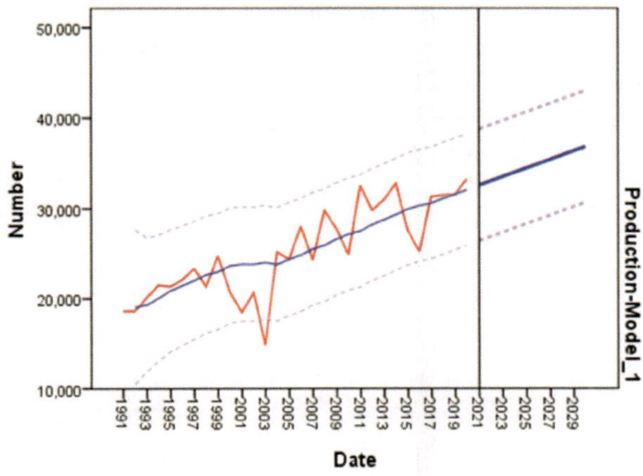

Fig.4.3. Series Chart with forecasted values of Productivity of Edible Oil Seeds in India from 1991 -92 to 2029-30

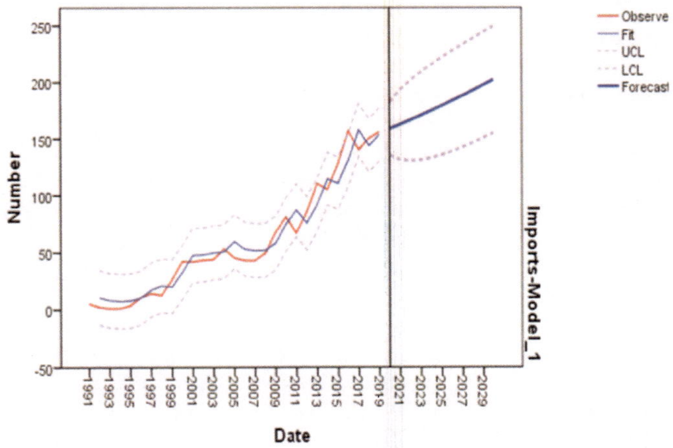

Fig.4.3. Series Chart with forecasted values of Productivity of Edible Oil Seeds in India from 1991 -92 to 2029-30

The above figure 4 obtained for the time series dataset for area cultivated (Fig.4.1), production (Fig 4.2) productivity (Fig.4.3) and import (Fig.4.4) from 1991-92 to 2029-30 are given in red colour and the predicted values for the same year given in blue colour. The red and blue lines are moving from left to right with very small deviations which tells the goodness of fit. The dotted lines are lower control limit and upper control limit of the variable. The fitted values are laying between these two controls limits. It also confirms that the constructed model is good. After the vertical line, there is bold blue line which is the line of predicted values from 2019-2020 to 2029-2030 and also lying between the control limits. There was moderate fluctuation in the movement of predicated lines for import observed which is moving towards upwards whereas area, production and productivity the fluctuation was not much high. Therefore increase in the cultivating area and productivity of edible oilseeds would helps to control the import of edible oil.

Conclusion

The study of forecasting import of edible oil is increasing even though increase in oilseeds area under cultivation, production and productivity. Since the land is limited factor production can be increased by rising yield i.e. productivity capacity of land through spreading inputs like technology in cultivation, high yield varieties of seeds, fertilizer and irrigation. Announcing the minimum support price during the sowing season would help to increase the cultivating area, production and productivity. These would help to reduce the import of edible oil in India.

References

Ajoy Kumar Singh, Anil Kumar Singh, Arvind Kumar Choudhary, Laradhan Kumari and Rakesh Kumar (2017). Towards Oilseeds Sufficiency in India: Present Status and way Forward. Journal of AgriSearch, Vol. 4(2), pp: 80-84.

Amin M., M. Amanullah and A. Akbar (2014), Time series modeling for forecasting wheat production of Pakistan. Plant Sciences. 24(5):1444-1451.

Ankur Jain (2018), Analysis of Growth and Instability in Area, Production, Yield and Price of Rice in India, Social Change and Development, Vol.XV, No.2, pp.49-66

Birhanu A and Sekar I (2016), Trends and regional Disparity of Maize Production in India Journal of Developpment and Agricultural Economics, Vo.8(9), pp.139-199, September.

Celik S , Karadas K and Eyduran E (2017)Forecasting the Production of Groundnut in Turkey using ARIMA Model ,The Journal of Animal & Plant Sciences, 27(3): 2017, Page: 920-928 ISSN: 1018-7081.

Celik, S. (2015).Modelling of honey production by using time series in Turkey. SAU Fen Bil. Der. 19 (2):377-382

Chandra Mohan Misra (2017), Trends in Area, Production and Productivity of Groundnut in India: Issues and Challenges, Quest Journals, Journals of Research in Agricultural and Animal Science, Vol.4, Issue 7, pp01-06, ISSN (online):2321-9459. www.questjournals.org.

Erik Dohlman et al.,(2003) Indian's Edible Oil Sector: Imports Fill Rising Demand, Electronic Outlook Report from the Economic Research Service. www.ers.usda.gov

Girish Kumar et al., (2012), Edible Oilseeds Supply and Demands in India: Implications for Policy, Division of agricultural Economics, Indian Agricultural Research Institute, New Delhi-110 012, pp 01-10.

GOI (Government of India): State of Indian Agriculture 2012-13, Ministry of Agriculture, New Delhi.

Hatje,G.(1989), World Importance of Oil Crops and their Products. In Oil Crops of the World, ed. G. Robbelen R.K. Downey, and A.Ashri,pp.1-21. New York: McGraw-Hill. https://www.worldatlas.com/articles/world-production-of-oilseeds-by-type.htm.

India Oilseeds: https://www.sopa.org/india-oilseeds-area-production-and-productivity/

John Beghin, Ndiame Diop, Holger Matthey (2006). Groundnut Trade Liberalization: Could the South Help the South? World Development Report, Vol.34, Issue 6, June, pp 1016-1036.

Narasinga Rao B.S (1992), Palm Oil as an edible oil in India and its role in meeting the nutritional needs of its population. Nutritional Research, Vol.12, supplement 1, pp:81-86.

Rajesh Bhayani (2020), Government Planning to Launch National Mission on Edible oil in the near future. Economic Policy, Business Standard, Mumbai.

Ramasamy C and Selvaraj K.N. (2002), Slow Growth Crops: Coarse cereals, Oilseeds and Pluses; Pulses, Oilseeds and Coarse cereals: Why they are slow Growth Crops? Indian Journal of Agricultural Economics Vol.57, No.3, July-September, pp 289-315.

Ravichandran S , Yashavanth B.S And Kareemulla K (2018), Oilseeds production and yield forecasting using ARIMA-ANN modelling, J. Oilseeds Res., 35(1) : 57-62, March, 2018, ICAR-National Academy of Agricultural Research Management, Rajendranagar, Hyderabad - 500 030, Telangana.

Sanjeev Kumar Redy (2020), Analysis of Oil Seed Crops Scenario in Rajasthan, Aayushi International Interdisciplinary Research Journal (AIIRJ), Vol. VII Issue.XI, November, : ISSN 2349-638x, PEER REVIEW e-JOURNAL; https://aiirjournal.com/uploads/Articles/2020/11/4792_07.DR.%20SANJEEV%20KUMAR%20BANSAL%20&%20DR.%20PARVEEN%20KUKKAR.pdf

Santosha Rathod, K N Singh, S G Patil, Ravindrakumar H Naik, Mrinmoy Ray And Vikram Singh Meena (2017), Modeling and forecasting of oilseed production of India through artificial intelligence techniques, Indian Journal of Agricultural Sciences 88 (1): 22–27, January 2018/Article , ICAR-Indian Agricultural Statistics Research Institute, New Delhi 110 012.

Size, Growth Rate an Distribution of population, pp.7. https://censusindia.gov.in/2011-prov-results/data_files/india/final_ppt_2011_chapter3.pdf

Stefansson (2015), Oilseed Crops, https://www.thecanadianencyclopedia.ca/en/article/oilseed-crops

Vijay Paul Sharma (2017), Oil Seeds Production in India: The Problems and Prospects, Springer, New Delhi, pp:3

World Production of Oilseeds:

25

Status and Performance of Crop Insurance in Himachal Pradesh

Amit, Niyati Thakur, Shilpa, Ajit Sharma and Harpreet Kaur

Department of Social Sciences, College of Forestry, Dr. Yashwant Singh Parmar University of Horticulture and Forestry, Nauni, Solan Himachal Pradesh, India

Introduction

India is an agricultural economy and the agriculture sector contributes 18.40 per cent to the Indian Gross Domestic Product (National Accounts Statistics, 2020-21). It is considered to be a major sector for its contribution to the Indian GDP. Agriculture is the source of income for 65 per cent population of the country. This sector is becoming innately insecure due to several reasons such as production, lack of financial resources, climate conditions and personal attitude of the young generation towards agriculture. It is right to say that agriculture is the "Gamble of Monsoon" because lack of precipitation can lead to drought and excessive precipitation can leads to floods. Around 55 per cent of arable lands depends on rainfall for irrigating the crops. After the 1960's green revolution, the dependence on HYV seeds, pesticides, fertilizers, etc. is increasing. The excessive use of the factors like pests affects the soil invertebrates and reduces their growth, pesticides secreted into the soil can end up in the ground-level water for drinking and the fertilizers leads adversely affect the fertility of the soil. The farmers are susceptible to agricultural risks because these factors are not only affecting the livelihood and income of the farmer but also tire the agricultural sector. Hence, threat management is one of the key concerns in agriculture for fixing the issues of poverty among farmers (Kumar et al., 2020). The complementary institutional tool is crop insurance which helps the farmers to alleviate financial losses due to different types of natural disasters (Prasad, 2018). It plays a significant role to stabilize the farmer's income against partial or complete crop failure due to adverse weather conditions such as floods, hail, drought and adverse physical crop conditions that are beyond control. Crop insurance enhances the farmer's investment in the agriculture sector and reduces the risk of crop damage by natural disasters (Jain, 2019).

Need of Crop Insurance

Both the Kharif and Rabi seasons had seen a decrease in crop yield due to extreme temperature and rainfall shocks. The productivity of agriculture is affected by this climatic change. In Indian agriculture, yield risk (variability in crop yield owing to weather) and price risk are two main types of risks. Despite the fact that they use more conventional risk management strategies, farmers diversify their crops to increase yields while reducing risk. In the absence of insurance arrangements, landless agricultural labourers and resource-poor farmers who have extremely limited means and resources are exposed. When natural disasters struck, the government provided relief packages as compensation, but these were severely constrained. In order to handle the issue of yield risk in the agriculture sector, crop insurance is required.

Benefits of Crop Insurance

When a farmer insures his crop, his mindset is modified (Ray *et al.,* 1997) and he is able to take agricultural risks that he would not have otherwise been able to (Ahsan *et al.,* 1982) because taking risks reduces capital investment in farming by inefficiently allocating resources (Jodha, 1981). As a result, having access to formal insurance enables farmers to reduce risk-related losses, maintain financial security, and boost farm revenue (Raju and Chand, 2007). The compensations paid in the event of crop failure aid the farmer in making on-time repayments on loans taken out through official lending institutions. Therefore, crop insurance raises the farmer's credit score by preventing loan default (Hazell et al., 1986; Pomareda, 1986 and Mishra 1996). Incentives for moral hazard brought on by insurance policies have occasionally caused insured farmers to use fewer chemical inputs, apply less nitrogen fertilizer, and affect maize producers' chemical use decisions (Horowitz and Lichtenberg, 1993). Therefore, a crop insurance program that is correctly designed and implemented prevents volatility, offers stability to farm revenues, and fills the gap between income and consumption needs during times of crop losses or crop failures (Bhende, 2002; Rajeev and Nagendran, 2019).

Evolution of Crop Insurance Schemes in India

In the early decades of the twentieth century, the concept of crop insurance has been derived and it's regarded as a significant instrument for risk management. J.S. Chakravarti worked between 1912-1920 and introduced a rain insurance scheme for the states of Mysore and the whole of India (Patidar, 2019). After the independence, the concept of crop insurance acquires more attention in India. The 'homogeneous area' approach for crop insurance was introduced by Prof. Dandekar in the mid-seventies. In 1972, the first crop insurance scheme was introduced. This experimental scheme was based on the "Individual Approach" which continued upto 1978. This scheme was working in the states like Gujarat, Tamil Nadu, Karnataka and Andhra Pradesh. It capped about 3110 farmers against claims of Rs. 37.88 lakh and a premium of amount Rs. 4.54 lakh (Jeyabalasingh P and Arvind N, 2020). After that, the General Insurance Corporation of India (GIC) develop a Pilot Crop Insurance Scheme (PCIS) which was implemented in 1979 and was based on the area approach. The scheme included grains, millets, cotton, oilseed, potatoes and grams. The ratio of risk between the state government and the GIC is 2:1(Yoga and Vetrival, 2012).

With the experience acquired from the introduction of the PCIS, the government of India with the collaboration of state governments implemented the Comprehensive Crop Insurance Scheme (CCIS) in April 1985. It is a short-term agricultural credit scheme and based on a homogeneous area. The central and state governments shared the premiums and claims in a ratio of 2:1 (Jeyabalasingh P and Arvind N, 2020). During the Rabi season 1999-2000,the new scheme was introduced i.e., National Agricultural Insurance Scheme (NAIS) by the Agriculture Insurance Company to replace the Comprehensive Crop Insurance Scheme. The primary purpose of thishomogeneous area-based approachis to save farmersand compensate and recover creditworthiness for crop failure losses from natural disasters such as floods, hailstorms, drought, cyclones, and fires (Mahajan and Bobde, 2012).

In 2007, the Weather Based Crop Insurance Scheme (WBCIS) was introduced in 20 states, to get more farmers under crop insurance homogenous area-based approach. Protecting the farmers from adverse weather conditions was the ultimate goal of the WBCIS scheme (Biswas et al., 2012). The Government of India implemented a new pilot project for Modified National Agricultural Insurance Scheme (MNAIS) in 50 districts from 2010-11 during the rabi season (Patidar, 2019). The key highlights of the scheme were on account of the settlement of claims, private firms with sufficient resources and expertise were permitted to enforce the scheme on the desired basis (Hazarika and Yasmin, 2018). For the years 2009-10 and 2010-11, the Coconut Palm Insurance Scheme (CPIS) was approved for pilot implementation in the selected areas of Andhra Pradesh, Goa, Karnataka, Kerala, Maharashtra and Odisha. CPIS was a mechanism designed to provide successful risk management support to all coconut growers who grow coconuts and are likely to be affected by natural phenomena such as storms, lightning, fire, earthquake, pests, infections and landslides (Rajasenan and Rajeev, 2013).

Current Status of crop insurance in India

The insurance scheme called Pradhan Mantri Fasal Bima Yojna (PMFBY) was launched in February 2016 by the Government. The theme of this scheme was **"One Nation-One Scheme"**(Singh and Singh, 2019). The major reason for launching this scheme at that timewas because the schemes viz., The National Agriculture Insurance Scheme (NAIS) and Modified NAIS (MNAIS) were not profiting the farmers. So, this scheme was introduced which covers all food crops, oilseeds crops and annual commercial/horticultural crops. The maximum premium payable by the farmers is 2 per cent, 1.5 per cent and 5 per cent for all Kharif crops, Rabi crops and annual commercial and horticultural crops respectively (Inderjeet, 2019). This scheme is mainly aimed at reducing agricultural distress and promoting farmers' welfare without having to affect hefty hikes in the minimum support prices (MSP) of agricultural products (Srinivasulu, 2015).

PMFBY aims to provide a comprehensive insurance cover against failure or damage of the crop thus helping in stabilizing the income of the farmers and encouraging them to adopt innovative practices. This is a crop insurance scheme for the farmers who are the backbone of the nation with the sole objective of supporting sustainable production in the agricultural sector by providing financial support to the farmers suffering from crop losses due to natural calamities. The scheme provides extended coverage under localized risks and post-harvest losses (Singh and Singh, 2019). The scheme aims at increasing crop insurance penetration in India through increased farmer awareness and low premium rates. PMFBY was revamped in 2020 and a major modification made under this scheme is now it has been made voluntary for all farmers earlier it was mandatory for loanee farmers and optional for non-loanee farmers (Anonymous, 2020).

Performance of PMFBY in India from 2016-17 to 2020-21

With the help of compound annual growth rate in indicators/components such as number of farmers covered, number of farmers benefitted, the area covered under the scheme (in hectares), total premium amount and total claims paid by insurance companies have been estimated to check the progress of PMFBY from 2016-17 to 2020-21. The number of farmers covered under PMFBY was 56.27 million in 2016-17 which increased to 60.62 million in 2020-21, the annual growth rate also showed an increasing trend (3.07 % p.a.). In 2016-17, the number of farmers who benefitted under PMFBY was 13.94 million which was increased to 14.42 million in the year 2020-21 and showed anannual growth of 3.38 per cent per annum. In 2016-17, the area covered under the scheme was 54.92 million hectares which decreased to 43.47 million hectares in 2020-21 and registered a decline at the rate of 4.66 per cent per annum. In the case of the total premium amount, it was 20.07 thousand crores in 2016-17 which increased to 28.61thousand crores in 2020-21. The compound annual growth rate recorded exceptionally high growth of 11.11 per cent per annum. Total claims paidto the farmers paid by insurance companies under PMFBY was15.15 thousand crores in 2016-17 which was increased to 23.07 thousand crores in 2020-21 and grew at the rate of 2.05 per cent per annum.

Table 1: Performance of PMFBY in India from 2016-17 to 2020-21

Indicators	2016-17	2017-18	2018-19	2019-20	2020-21	CAGR (%)
No. of Farmers covered (in millions)	56.27	51.24	55.63	59.76	60.62	3.07
No. of Farmers Benefitted (in millions)	13.94	16.14	20.53	21.04	14.42	3.38
Area Covered (in millions ha.)	54.92	48.75	50.70	48.30	43.47	-4.66
Total Premium amount ('000 Crores)	20.07	22.24	26.39	28.61	29.96	11.11
Total Claims Paid ('000 Crores)	15.15	20.25	24.80	23.07	15.71	2.05

Source: Department of Agriculture & Farmers Welfare

Status of PMFBY in Himachal Pradesh

Himachal Pradesh is a north-western Himalayan state. The ranges of climate from subtropical to subarctic cold with an annual rainfall of 350-3800 mm. Its temperature varies from -25°C to 42°C. Himachal Pradesh has a geographical advantage with a wide range of topography, soil, climate, rainfall and temperature conditions that provide wide opportunities and potential for agriculture and horticulture. The farmers in Himachal are essentially involved in mixed farming, agro-pastoral, Silvi-pastoral and agro-horticultural systems because of their topography. In the total Gross State Domestic Product (GSDP), the agriculture sector of Himachal state contributes nearly about 13.31 per centand employs about 69 per cent of the total workers of the State (Economic Survey of Himachal Pradesh, 2020). Still,agriculture faces several risks, mainly in the production of food grains, natural disasters and uncertainty faced by the farmers in Himachal Pradesh.The Government of India launched its flagship scheme Pradhan Mantri Fasal Bima Yojana to reduce the risk of crop failure. PMFBY is a large-scale crop subsidy insurance scheme that provides risk cover to Indian farmers from production susceptibility. In 10 districts of the state except for the Lahaul & Spiti and Kinnaur districts has been implemented the scheme. In 2016 the different agencies viz., Agriculture Insurance Company (AIC), IFFCO-TOKIO and Oriental Insurance Company (OIC) have been implemented the PMFBY.

The crops like wheat, maize, barley, seed-potato, vegetables and mushrooms, etc are the major crops that are grown in Himachal Pradesh.Apart from them, apples, plums, peaches, apricot, mango and citrus, etc. are major fruit crops. Crop failure occurs every year in Himachal Pradesh due to terrain and natural disaster. So, crop insurance is the only method to save the production losses of farmers in Himachal Pradesh. To keep these things in mind, the government introduced the scheme i.e., Pradhan Mantri Fasal Bima Yojana, in 10 districts of the state except Kinnaur and Lahaul and Spiti in 2016 during the Kharif season (Kalia et al., 2018). The farmers covered was 89,662 with an area of 33,743 ha insuredduring Rabi in 2019-20and the total premium paid was Rs. 9 crores along with a claim received amount of Rs. 5.48 crore. In 2020 during the Kharif season, 85,492 farmers were covered under this scheme with an area of 30,593 ha and the total amount of premium paid was 7.79 crore. During Rabi 2019-20, there were two implementing agencies viz. The New India Assurance Company and Agriculture Insurance Company (AIC) during the Rabi season in 2019-20, however only one executing agency was Agriculture Insurance Company in 2020 Kharif season (Anonymous, 2019).

Progress of PMFBY in Himachal Pradesh from 2016-17 to 2020-21

To figure out the performance of the scheme from 2016-17 to 2020-21, the two indicators such as percentage change and compound annual growth rates have also been estimated.From the table, it can be observed that by adopting the PMFBY, remarkable growth has been seen in Himachal Pradesh.Initially, the farmers covered under PMFBY was2.61 lakhs in 2016-17which increased to 10.03 lakhs in 2020-21 and recorded a growth of 38.03 per cent per annum. The farmers who benefitted from PMFBY were0.23 lakhs in 2016-17 which increased to 2.89 lakhs in 2020-21 and registered high growth of 79.69 per cent per annum. The area covered under PMFBY was 0.85 lakh ha in 2016-17 which was increased to3.55 lakh ha in 2020-21 and grew at the rate of 41.04 per cent per annum.Over the period, the total premium amount was also found to increase, 1.54 thousand lakhs during 2016-17 which was increased to 7.05 thousand lakhs and recorded a growth of 43.58 per cent per annum. Total claims paid was 0.57 thousand lakhs which was increased to 5.45 thousand lakhs and registered 69.12 per cent growth annually.

Table 2: Progress of PMFBY in Himachal Pradesh from 2016-17 to 2020-21

Indicators	2016-17	2017-18	2018-19	2019-20	2020-21	CAGR (%)
No. of Farmers covered (in millions)	2.61	4.91	6.48	8.32	10.03	38.03
No. of Farmers Benefitted (in millions)	0.23	0.86	1.23	1.93	2.89	79.69
Area Covered (in millions ha.)	0.85	1.63	2.26	2.93	3.55	41.04
Total Premium amount ('000 Crores)	1.54	2.84	3.69	5.08	7.05	43.58
Total Claims Paid ('000 Crores)	0.57	1.72	2.56	3.59	5.45	69.12

Source: Department of Agriculture & Farmers Welfare.

References

Agricultural Statistics at a Glance 2020-21. Department of Agriculture & Farmers Welfare. Ministry of Agriculture & Farmers Welfare, Krishi Bhawan, New Delhi. 431p

Ahsan, Sayed M, Ali AG and Kurian J. 1982. Towards a theory of Agricultural Insurance, American Journal of Agricultural Economics 64: 510-529. Available at: https://doi.org/10.2307/1240644.

Anonymous. 2019. Department of Agriculture, Shimla, Himachal Pradesh.www.hpagriculture.com(1.00PM, 10th February 2021)

Anonymous. 2020. Revamped operational guidelines of Pradhan Mantri Fasal Bima Yojana. https: pmfby.gov.in (5.00 PM, 14March, 2021)

Bhende MJ. 2002. An analysis of Crop Insurance Scheme in Karnataka. Bangalore: Agricultural Development and Rural Transformation Unit, Institute for Social and Economic Change (ISEC).

Biswas B, Dhaliwal L K, Singh S P. and Chahal S K. 2009. Weather-Based Crop Insurance in India: Present status and future possibilities. Journal of Agrometeorology 11:238-41.

Economic Survey (2020). Department of Economics and Statistics, Government of Himachal Pradesh, Shimla. 161p.

Hazarika C and Yasmin S. 2018. Adaptability of crop insurance as a risk mitigation mechanism by the farmers of Assam - An analysis of Modified National Agricultural Insurance Scheme (MNAIS). Journal of Advanced Agricultural Technologies5:58-62.

Hazell P, Bsssoco LM and Arcia G. 1986. A Model for Evaluating Farmers" Demand for Insurance : Applications in Mexico and Panama. In P.B.R.Hazell, C.Pomareda. and A.Valdes (eds), Crop Insurance for Agricultural Development : Issues and Experience. Baltimore and London: The Johns Hopkins University Press.

Horowitz JK and Lichtenberg E. 1993. Insurance, Moral Hazard, and Chemical Use in Agriculture. American Journal of Agricultural Economics 75: 926-935.

Inderjeet. 2016. Study on the benefits of Pradhan Mantri Fasal Bima Yojna for farmers. International Journal of Multidisciplinary Research and Development6:384-87.

Jain N. 2019. Perception of farmers towards Pradhan Mantri Fasal Bima Yojana (PMFBY) at Sehore block of Sehore District (M.P.). M.Sc. Thesis. Department of Agricultural Extension, Rajmata Vijayaraje Scindia Krishi Vishwa Vidyalaya, Gwalior. 92p.

Jeyabalasingh P and Arvind N. 2020. Crop Insurance in India: Evolution, issues and way forward. Journal of Critical Reviews7:2851-69.

Jodha NS. 1981. Role of Credit in Farmers" Adjustment against Risk in Arid and Semi-Arid Tropical Areas of India. Economic and Political Weekly 16: 1696-1709. Available at: https://www.epw.in/system/files/pdf./1981_16/special_articles_role_of_credit_in_farmers_adjustment.pdf

Kalia A, Sadrate N D, Darhel V. and Singh R. 2018. Performance evaluation of Pradhan Mantri Fasal Bima Yojana (PMFBY) in Himachal Pradesh. Agro-Economic Research Centre Himachal Pradesh University, Shimla. 142p.

Kumar S, Sharma A. and Kumar S. 2020. Pradhan Mantri Fasal Bima Yojana (PMFBY): A tool for agricultural risk management.Indian Farmer7:467-71.

Mahajan S S and Bobde A H. 2012. Growth of NAIS: A study of crop insurance in India. Buaddhik3:1-15.

Mishra PK.1996. Agricultural Risk, Insurance and Income. Arabury, Vermont: Ashgate Publishing Company.

National Accounts Statistics (2021). Central Statistical Organisation. Ministry of Statistics and Programme Implementation, Government of India. pp 408.

Patidar N K. 2019.Knowledge and opinion of farmers towards Pradhan Mantri Fasal Bima Yojana in Jaipur district of Rajasthan (SKNAU). M.Sc. Thesis. Department of Extension Education, Sri Karan Narendra Agriculture University, Jobner. 167p.

Pomareda C. 1986. An Evaluation of the Impact of Credit Insurance on Bank Performance in Panama. In Peter Hazell, Carlos Pomareda and Alberto Valdes. (eds), Crop Insurance for Agricultural Development : Issues and Experience. Batimore and London: The Johns Hopkins University Press.

Prasad D. 2018. Assessment of the Pradhan Mantri Fasal Bima Yojana (PMFBY) with reference to satisfaction among the rice growers in Mahasamund district of Chhattisgarh. M.Sc. Thesis. Department of Agricultural Economics, Indira Gandhi Krishi Vishwavidyalaya, Raipur. 87p.

Rajasenan D and Rajeev B. 2013. Stakeholder's perception of Coconut Palm Insurance Scheme: A study on the coastal states of India. Cord 29:1-11.

Rajeev M and Nagendran P. 2019. Where Do We Stand? Crop Insurance in India, Economic and Political Weekly 54:28-35. Available at: https://www.epw.in/system/files/pdf./2019_54/27/SA_LIV_26%2627_290619_RRA_Meenakshi_Rajeev.pdf.

Raju S and Chand R. 2007. Progress and Problems in Agricultural Insurance in India, Economic and Political Weekly 42: 1905-1908. Available at: https://www.epw.in/system/files/pdf./2007_42/21/Progress_and_Problems_in_Agricultural_Insurance.pdf.

Ray PK. 1971. Review of Paddy Crop Insurance in Ceylon. Colombo: Government of Ceylon.

Singh A K and Singh A. 2019. Pradhan Mantri Fasal Bima Yojna (PMFBY): A bunch of Benefits. International Journal of Multidisciplinary 4:1031-35

Srinivasulu M. 2015. Agriculture crop insurance policies in India - A study on Pradhan Mantri Fasal Bima Yojana (PMFBY) in Telangana state. International Journal of Innovative Research in Engineering & Multidisciplinary Physical Sciences 3:13-18.

Vani G. 2019. A review on Pradhan Mantri Fasal Bima Yojana. International Journal of Scientific Research and Review 7:3465-71.

Yoga S and Vetrival K 2012. Origin and progress of crop insurance in India - A historical view. Review of Research1:1-4.

26

Biodiversity Conservation

Veena Sharma[1] Mahender Singh[2] and Vikas Sharma[3]

Division of Soil Science and Agricultural Chemistry, SKUAST-Jammu Chatha, Jammu and Kashmir, India

Biodiversity

Biological diversity, or biodiversity, is the scientific term for the variety of life on Earth. It refers not just to species but also to ecosystems and differences in genes within a single species. Everywhere on the planet, species live together and depend on one another. Every living thing, including man, is involved in these complex networks of interdependent relationships, which are called ecosystems. Healthy ecosystems clean our water, purify our air, maintain our soil, regulate the climate, recycle nutrients and provide us with food. They provide raw materials and resources for medicines and other purposes. They are at the foundation of all civilisation and sustain our economies. It's that simple: we could not live without these "ecosystem services". They are what we call our natural capital.

Biodiversity is the key indicator of the health of an ecosystem. A wide variety of species will cope better with threats than a limited number of them in large populations. Even if certain species are affected by pollution, climate change or human activities, the ecosystem as a whole may adapt and survive. But the extinction of a species may have unforeseen impacts, sometimes snowballing into the destruction of entire ecosystems. Biodiversity refers to the various life forms that exist on earth, including animals, plants, microorganisms, and the entire ecosystem they live in. Biodiversity is in the form of biological resources, including genes, species, organisms, and ecosystems. Thus, the four main levels of biodiversity are species, genetic, ecosystem, and global biodiversity.

Because of these characteristics, biodiversity plays a fundamental role in maintaining the aesthetic value of the environment, the integrity of the natural environment, and promotes the overall well-being of all plant and animal life. This calls for the need for biodiversity conservation for the survival of all living things and their natural habitat. Hence, biodiversity conservation is all about protecting all organisms and species within their natural habitats with the aim of ensuring intragenerational and intergenerational equity. Activities such as habitat fragmentation, human disturbance, and habitat loss have to be adequately curtailed to enrich biodiversity conservation measures.

Threats to Biodiversity

Over the last century, humans have come to dominate the planet, causing rapid ecosystem change and massive loss of biodiversity across the planet. This has led some people to refer to the time we now live in as the "anthropocene." While the Earth has always experienced changes and extinctions, today they are occurring at an unprecedented rate. Major direct threats to biodiversity include habitat loss and fragmentation, unsustainable resource use, invasive species, pollution, and global climate change. The underlying causes of biodiversity loss, such as a growing human population and overconsumption are often complex and stem from many interrelated factors.

Causes of Biodiversity Depletion

Among these causes, habitat destruction and over-exploitation are the main causes for this catastrophe. Habitat (natural home) destruction: may result from clearing and burning forests, draining and filling of wetlands, converting natural areas for agricultural or industrial uses, human settlements, mines, building of roads and other developmental projects. This way the natural habitats of organisms are changed or destroyed. These change either kill or force out many species from the area causing disruption of interactions among the species. Over exploitation reduces the size of the population of a species and may push it towards extinction.

Overall, the main factor directly driving biodiversity loss worldwide is habitat alteration and destruction. Habitat destruction renders entire habitats functionally unable to support the species present in the habitat. Biodiversity reduced in this process when existing organisms in the habitat are displaced or destroyed (Ayoade at al., 2009; Agarwal et al., 2011).Introduction of exotic species: Exotic species are new species entering geographical region. Seeds catch on people's clothes. Mice, rats and birds hitch-hike on ships. When such species land in new places, they breed extra fast due to absence of any enemy and often wipe out the native species already present there. Exotic species may wipe out the native ones. A few examples are- Water hyacinth which clogs lakes and riversides and threatens the survival of many aquatic species. This is common in Indian plains. Similarly Lantana camara (an American weed) has invaded many forest lands in various parts of India and wiped out the native grass species. Apart from these examples there are several others that have caused disruption of natural ecosystems in India as well as abroad.

Pollution: Air pollution, acid rain destroy forests. Water pollution kills fishes and other aquatic plants and animals. Toxic and hazardous substances drained into waterways kill aquatic life. Oil spills kill coastal birds, plants and other marine animals. Plastic trash entangles wildlife. It is easy to see how pollution is a big threat to biodiversity. The vulture was once very common in the Gangetic plains of India, and often seen nesting on the avenue trees within large cities in the region. Before the 1990s they were even seen as a nuisance, particularly to aircraft as they were often involved in bird strikes. The vulture has suffered a 99% population decrease in India (Prakash, 2007) and become rare due to poisoning by DDT used as pesticides and also by diclofenac which is used as veterinary non –steroidal anti-inflammatory drug, leaving traces in cattle carcasses which when fed by vultures leads to thinning of egg shells resulting into premature hatching and kidney failure in birds (Green et al., 2004;Muralidharan et al., 2008).

Population growth and poverty: Over six billion people live on the earth. Each year, 90 million more people are added. All these people use natural resources for food, water, medicine, clothes, shelter and fuel. Need of the poor and often greed of the rich generate continuous pressure resulting in over-exploitation and loss of biodiversity.

Types of Biodiversity Conservation

1. Genetic Diversity

Genetic diversity encompasses all the variety of different genetic information contained in animals, microorganisms and plants. A single species of organisms with diverse genetic diversity portray more adaptability and survival mechanism against adverse environmental conditions compared to organisms of a single species with the same genetic make-up. As per UNEP estimates there are 9.0 to 52 million species exist on earth (Mora et al., 2011).

2. Species Diversity

Species diversity refers to the variety of organisms in the environment. They can be further subdivided into:

- **Species richness**. It is the total number of species in a given region or locality. It is measured using the Menhink and Mangalet indices.
- **Special abundance**. This is the relative number of organisms among the species (the number of individuals per species). When species have equal abundance, their variation becomes high, which also makes the diversity to be high. A low diversity scenario is noted when the number of one species is 98 and the other species 1.
- **Taxonomic diversity (Phylogenetic)**. Taxonomic diversity (TD) is the genetic relationship that exists between various groups of species. This kind of diversity is represented using a hierarchical classification based on the phylogenetic evolution of different species of organisms.

3. Ecosystem Diversity

It is relatively similar to species diversity except for the fact that it deals with the variations in ecosystems within a geographical location. From one region or country to another, there are different ecosystems or biomes with examples such as alpine meadows, taigas, wetlands, grasslands, mangroves, and deserts, among others.

Biodiversity is not distributed evenly on Earth. It is the richest in the tropics. Terrestrial biodiversity tends to be highest near the equator, which seems to be the result of the warm climate and high primary productivity (Field et al., 2009). Marine biodiversity tends to be highest along coasts in the Western Pacific, where sea surface temperature is highest and in the mid-latitudinal band in all oceans. There are latitudinal gradient in species diversity. Biodiversity generally tends to cluster in hotspots, and has been increasing through time (McPeek et al., 2007) but will be likely to slow in the future (Robosky, 2009).

4. Global Diversity

Global diversity refers to the range of differences that depict the composition of a group of two or more species in a global context. According to the International Union for Conservation of Nature (IUCN), there is approximately 1. 5 million different species of plants and animals spread across the world. 70% represent animal species, while 22% represent plants.

Importance of Biodiversity Conservation

1. Biodiversity Conservation is Important for Economic Growth and Poverty Reduction

This argument is based on the numerous direct economic benefits that humans derive from natural environments such as food, construction material, fiber, firewood, industrial products, and medicinal value. As an example, approximately 25,000 plant species are used by native people to extract traditional medicine. Plus, more than 25% of drugs sold worldwide are derived from plants. And according to the UN, the majority of the world's poor lives in rural areas and depend upon wetlands, forests, pastures, and water for their livelihoods.

Billions of people worldwide also depend on timber products for income and subsistence. The national parks and sanctuaries are a source of tourism that amaze people as a source of natural beauty.

2. Supports the Continuity of Various Ecosystems Globally

Biodiversity plays an essential role in ecosystem rejuvenation and protection. The Amazon forest, for example, can produce approximately 20% of the total oxygen on earth through photosynthesis. There are also thousands of pollinators such as insects and birds, and other numerous biological activities that take place in the forest. If the biodiversity of the forest is conserved, it means the entire ecosystem that relies on the Amazon forest is equally protected and allowed to rejuvenate. Other ecosystems like the coral reefs, tundra, rivers and streams, and grasslands can as well be supported and protected through biodiversity conservation.

3. Aesthetic Value

The natural environment provides great pleasure to human beings with its shape, structure, senses, and color, which enriches people's culture. Activities such as visits to animal parks, bird watching, nature art, and cultural heritage as ways of enjoying and appreciating nature are only made possible through biodiversity conservation.

4. Ecological Balance

The integrity of the ecology – the harmonious coexistence of organisms and their environment is preserved by biodiversity. Some of the main aspects can be explained through:

- Carbon dioxide and oxygen balance (which helps in addressing the effects of climate change). The sequential balance between atmospheric carbon dioxide and oxygen are maintained through biodiversity. Failure to conserve biodiversity leads to the accumulation of carbon dioxide resulting in a greenhouse effect and the gradual depletion of ozone. The results are global warming and natural calamities.
- Biochemical cycles. An example is a hydrological cycle. The availability of biological resources such as forests and wetlands are vital for biochemical cycles. Lack of these resources would lead to incomplete cycles and increased incidences of natural calamities like desertification and species extinction.
- Decomposition. Biodiversity conservation protects decomposers – organisms that aid in the breakdown of waste organic matter. Decomposition is thus an essential part of the food chain as it transforms the waste of dead organic matter, which is then converted into other biochemicals that are availed for primary producers.
- Climate. The maintenance of the micro, local or regional climate is determined and regulated by biodiversity. Biodiversity achieves this by influencing the air turbulence, temperature, and precipitation.

5. Ethical Value for Every form of Life in the Environment

Ecosystem's right of an organism states that every form of life in any ecosystem is unique and deserves respect from human beings. The right suggests that every organism on earth, whether it is valuable to humans or not, has an inherent right to exist. Besides, the present generation has a social responsibility towards future generations, and this includes the protection of all living organisms in the world. Through biodiversity conservation, therefore, the ethics on environmental sustainability and conservation can be encouraged.

6. Ecosystem Services

Ecosystem services mean processes the biodiversity provides to support human life. For example, pollination, decomposition of waste, water purification, renewal of soil fertility and moderation of floods. Ecosystem services can be separated into three categories: Provisioning services include anything related to the production of renewable resources, like farming or energy production. Regulating services mean anything that lessens environmental change. Cultural services are anything that provides direct value or enjoyment from natural resources and beauty.

Ecosystem processes are not generally valued as part of the economy until they cease to function. When the economic value is assigned to these services, they are worth billions of dollars. For example, insect pollinators help in the production of many commercially important fruits such as almonds, melons, blueberries, and apples. Similarly, in other ecosystem services, water purification just involves filtering of rainwater by soil and by microbes that can break down nutrients and contaminants and reduce metal ions, slowing their spread into the environment. Wetland and riparian plants absorb nitrogen and trap sediments that decrease water quality.

As human construction and development disrupt natural environments as well as all activity and services related to this environment, finally, we have to be dependent on artificial man-made services like for filtration of water using filters and purifiers. These artificial services cost much more while the natural ecosystem services are free of cost.

7. Social and Spiritual Benefits

In human history, conservation also means protecting nature for its spiritual gifts and protecting sacred places in the local landscape. The biodiversity effects on cultural development can be shown by the heterogeneity of the world's mythology, folk dances and folk art, which contribute to the richness of literature and global arts. Different landscapes with different cultures influence our language, diet, occupation and various types of activity.

The uniqueness of each habitat consists of animals and plants present in each country and state with their flagship animals as well as plants. Even during traveling, the sight of biological diversity, different cultures and landscapes motivate peoples. Ecotourism is traveling with the aim to view, support and sustain the local cultures and its natural ecosystem. Support from ecotourism can be very helpful to reduce habitat destruction as well as to preserve endangered species.

Methods of Biodiversity Conservation

The process of conservation of biodiversity involves sustainable development. It also involves the *in situ* conservation where conserving and protecting the ecosystem will protect the entire biodiversity.

At the same time, it includes *ex-situ* conservation, which involves conservation measures when an organism is endangered. Therefore, this means that the conservation methods of biodiversity aim at preservation, maintenance, conservation, recovery, and enhancement as discussed below:

1. Protection Against Degradation and Destruction of Natural Ecosystems

Biodiversity conservation involves promoting a balance between the environment, society, and development in line with the sustainable development goals. The present society should develop conservation strategies that meet their need without compromising that of the future generation. Having a balance between the development of the environment and society results in ensuring the achievement of biodiversity conservation. Through properly enforced policies, sustainable development can be achieved. Environmental institutions can enforce these conventions.

2. Maintain, Restore, and Increase Ecological Systems While Promoting the Implementation of Better Conservation Practices

It involves the in-situ and ex-situ methods. In situ conservation involves the conservation of the whole ecosystem and the natural habitats. It also includes the maintenance and recovery of a variety of species and degraded ecological systems such as damaged forest areas and heavily polluted lakes, rivers, and lands, among other natural environments. The ex-situ conservation involves the strategic protection of biodiversity hotspots for endangered species and habitats.

3. Identification and Protection of Endangered Species

Ex-situ biodiversity conservation aims to reduce the ongoing extinction up to 30% or higher by mainly focusing on two things. First, is the focus on conservation of biodiversity in their natural habitats in places like museums, arboretums, zoos, and gene banks. High biodiversity areas should be covered in the form of natural parks, sanctuary, biosphere reserve etc.

The second strategy focuses on captive conservation methods that involve the protection of endangered species. This type of conservation is a collective responsibility of all nations. An example is the Earth Summit, which was held in Rio de Janeiro, asking nations to take appropriate measures towards the protection of vulnerable habitats and species.

4. Establishing Buffer Zones to Prevent any Alteration in the Balance of Natural Ecosystems

Biodiversity conservation prioritizes on establishing control measures that maintain the balance of natural ecosystems such as water balance, soil ecology, and genetic as well as species balance. It involves intragenerational and intergenerational equity that allows equitable sharing of resources and the benefits derived from natural habitats to ensure social and ecological stability. Some of the conservation efforts that have been advanced under this category include the World Conservation Union, International Board for Plant Genetic Resources, and UNESCO program on Man and biosphere.

5. Science, Technology and Research-Tools for Conserving Biodiversity

As our society develops, it leads to improved science and technology. Science, and specifically ecology, help scientists to understand the web of interactions in our biomes and pinpoint the key species in ecosystems. This information is used to guide conservation efforts. The same is also used to understand pollution and its cascading effects within an ecosystem as Bio-magnification of toxins in a food chain can cause huge problems for top predators.

Technology is becoming vital in conservation biology. Sustainable technologies, like renewable energies, biodegradable packaging, and recycling, help reduce human impact on the environment. There are also technologies like cloning that give scientists the ability to bring back species that are already considered extinct. Research on how species interact within their environment is crucial for the protection of organisms and maintaining biodiversity. For example, the use of wildlife corridors in urbanized areas, forming zoo and botanical gardens for research purposes and collections of living organisms, In vitro plant tissue and microbial culture dramatically increases their populations.

6. Captive Breeding and Gene Banks

When animals are bred in captivity (often at zoos), it is called captive breeding. It requires the capture of animals that are often near extinction. Captive breeding of animals and artificial propagation of plants is somewhat controversial; however, on the positive side, it provides the opportunity to increase the population of the species so that they can be reintroduced into the wild. Gene banks store seeds, sperm & ova at extremely low temperatures and humidity. It is very helpful to save a large variety of species of plants & animals in a very small space such as sperm and ova banks, seedbanks. Seed banks are areas where huge varieties of plant seeds are stored. Seed banks have been collecting samples for many years, with some having over 2 billion seeds stored at a time. If a species goes extinct in nature, it provides a failsafe. The plant can be grown from saved seed and reintroduced back into its habitat.

Conclusion

It is imperative that the phenomenon of biodiversity is very vast, complex and interdependent and there is no single over-arching effect of diversity on either productivity or stability. The realized effects will depend heavily on environmental context and the time scale over which the effects are studied. t is therefore necessary for legislators to understand the basic science in order to maintain diversity at its current levels. If current human growth and resource management patterns do not change, it is likely that we will lose many important species, and the ecosystems of the world may never recover.

References

Agarwal, N.K., 2011. Cryopreservation of Fish Semen In. J.P. Bhatt, Madhu Thapliyal and Ashish Thapliyal (eds.), Himalayan Aquatic Biodiversity Conservation & New Tools in Biotechnology, Transmedia Publication, Srinagar (Garhawal) Uttarakhand. pp: 104-127

Ayoade, A. A., Agarwal, N.K and Chandola-Saklani, A., 2009. Changes in Physico-chemical Features and Plankton of Two Regulated High Altitude Rivers, Garhwal Himalaya, India. European Journal of Scientific Research, 27 (1): 77-92

Field, R., Hawkins, B. A., Cornell, H.V., Currie, D.J., Diniz- Filho, J., Alexandre, F., Guégan, Jean-François; Kaufman, D.M., Kerr, J.T., Mittelbach, G.G., Oberdorff, T., O'Brien, E. M. and Turner, J. R. G., 2009. Spatial species-richness gradients across scales: a meta-analysis. Journal of Biogeography, 36 (1): 132– 147. doi:10.1111/j.1365-2699. 2008. 01963.x

Green, R. E., Newton, I., Shultz, S., Cunningham, A. A., Gilbert, M., Pain, D. J. and Prakash, V., 2004. Diclofenac poisoning as a cause of vulture population declines across the Indian subcontinent. Journal of Applied Ecology, 41(5): 793–800. doi:10.1111/j.0021-8901. 2004. 00954.

McPeek, M.A. and Brown, J.M., 2007. Clade Age and Not Diversification Rate Explains Species Richness among Animal Taxa. The American Naturalist, 169 (4): E97–E106. doi:10.1086/512135. PMID 17427118

Mora, C., Tittensor, D.P., Adl, S., Simpson, A.G. and Worm, B., 2011. How many species are there on Earth and in the ocean?. PLOS Biology. 9(8): e1001127. doi:10.1371/journal. pbio.1001127. Retrieved on 26 May 2015.

Muralidharan, S., Dhananjayan, V., Risebrough, R., Prakash, V., Jayakumar, R. and Bloom, Peter H., 2008. Persistent Organochlorine Pesticide Residues in Tissues and Eggs of White-Backed Vulture, Gyps bengalensis from Different Locations in India. Bulletin of Environmental Contamination and Toxicology, 81 (6): 561–565. doi:10.1007/s00128-008- 9529-z. 18806909.

Prakash,V.,2007.RecentchangesinpopulationsofresidentGypsvulturesinIndia.JournalofBombay Natural History Society, 104 (2): 129–135

Rabosky, D.L., 2009. Ecological limits and diversification rate:alternative paradigms to explain the variation in species richness among clades and regions. Ecology Letters, 12 (8): 735–743. doi:10.1111/j.1461- 0248.2009.01333.x. PMID 19558515.